Technology and Regulation

2019
Volume 1

OPEN PRESS TiU

TECHNOLOGY AND REGULATION 2019

Volume 1

DOI: 10.26116/techreg.volume.2019
ISBN: 9789403662411

Technology and Regulation
Tilburg Institute for Law, Technology, and Society (TILT)
Tilburg Law School
P.O. Box 90153
5000 LE Tilburg
The Netherlands
techreg.org

Principal Contact:
Ronald Leenes
Editor-in-Chief
Tilburg Institute for Law, Technology, and Society (TILT), Tilburg Law School
r.e.leenes@tilburguniversity.edu

Support Contact:
Aaron Martin
a.k.martin@uvt.nl

Published by: Open Press TiU
Contact details: info@openpresstiu.edu
https://www.openpresstiu.org/

Cover Design by: Wolf Publishers, Claudia Tofan & KaftWerk/Janine Hendriks

Open Press TiU is the academic Open Access publishing house for Tilburg University and beyond. As part of the Open Science Action Plan of Tilburg University, Open Press TiU aims to accelerate Open Access in scholarly book publishing.

OPEN PRESS Tilburg University 2021

Aims and Scope

Technology and Regulation (TechReg) is an international journal of law, technology and society, with an interdisciplinary identity. TechReg provides an online platform for disseminating original research on the legal and regulatory challenges posed by existing and emerging technologies (and their applications) including, but by no means limited to, the Internet and digital technology, artificial intelligence and machine learning, robotics, neurotechnology, nanotechnology, biotechnology, energy and climate change technology, and health and food technology. We conceive of regulation broadly to encompass ways of dealing with, ordering and understanding technologies and their consequences, such as through legal regulation, competition, social norms and standards, and technology design (or in Lessig's terms: law, market, norms and architecture). We aim to address critical and sometimes controversial questions such as: How do new technologies shape society both positively and negatively? Should technology development be steered towards societal goals, and if so, which goals and how? What are the benefits and dangers of regulating human behaviour through technology? What is the most appropriate response to technological innovation, in general or in particular cases? It is in this sense that TechReg is intrinsically interdisciplinary: we believe that legal and regulatory debates on technology are inextricable from societal, political and economic concerns, and that therefore technology regulation requires a multidisciplinary, integrated approach. Through a combination of monodisciplinary, multidisciplinary and interdisciplinary articles, the journal aims to contribute to an integrated vision of law, technology and society. We invite original, well-researched and methodologically rigorous submissions from academics and practitioners, including policy makers, on a wide range of research areas such as privacy and data protection, security, surveillance, cybercrime, intellectual property, innovation, competition, governance, risk, ethics, media and data studies, and others.

TechReg is double-blind peer-reviewed and completely open access for both authors and readers. TechReg does not charge article processing fees.

Editorial Team

CONTENTS

01

Technology and Regulation

Of Horses and Other Animals of Cyberspace
Editorial

Ronald Leenes*

Technology regulation, technology law, cyber-law

r.e.leenes@tilburguniversity.edu

In this introductory article to the new journal *Technology and Regulation,* I give a somewhat personal account of the history of cyberlaw and technology law and the 'struggles' some scholars have finding their spot in the more general legal realm. It will recount some of the classic discussions in the field, such as whether cyberlaw is just a form of the 'Law of the Horse'. It also outlines the contours of the field of technology regulation, some of the open questions in defining this field and some of its constituent elements. Finally, questions that I hope will be addressed in future articles in the journal are provided.

1. Coping with change

Courts are used to coping with change. Sometimes they face changes in society, for instance due to technological advances, which shake the foundations on which legal concepts are grounded. The move from atoms to bits is an example of such foundational friction. Courts and legislators have had to deal with questions about how the law relating to atoms applies to cases involving bits (in the absence of bits-specific law), facing the fact that bits and atoms have different properties and finding that the law is not adequately suited to cope with relevant differences.

In the Dutch legal history, a long legal battle was fought about the proper legal treatment of certain intangibles. It started in 1921 with a dentist in The Hague tapping electricity from the grid by tampering with his electricity meter. The courts, up to the Supreme Court, faced the question whether this act amounted to electricity theft. The Criminal Code at the time was tailored to deal with tangible objects, as was the Civil Code. It talks about taking away 'goods', which is traditionally understood as physically taking something in one's hands and running off with it. The Dutch Supreme Court adopted a teleological interpretation of the provision, stating that its purpose is to protect the assets of its owner. Assets generally have some independent existence, can be controlled by humans, can be transferred and accumulated and represent a certain value, according to the Court. Electricity shares these properties, and – like tangibles but unlike intellectual property – is the product of physical labor (they are atoms rather than bits), and can hence be seen as an asset that is worthy of protection against theft, the ruling states.[1] With this ruling, goods lost their tangibility under Dutch criminal law. Legal scholarship was divided over the extensive interpretation of the concept of 'good' adopted by the court, which was deemed infringing the Nulla poena sine lege stricta[2] principle.

A next case in this series concerned a woman who had accidently received a sum of money on her bank account. She subsequently spent the money, but was charged with embezzlement (art. 321, Dutch Criminal Code). Following the 1921 electricity reasoning, the Supreme Court qualified credit on a bank account under 'good' as mentioned in the Criminal Code because the credit represents value and furthermore the money can be spent only once.[3] Thus, cashless money – which consists of bits rather than atoms – was brought under the concept of 'good'.

This raised questions when computer data, as a new species of intangibles, came up in cases in which defendants were charged with theft or embezzlement. Initially, various Dutch courts adjudicated cases concerning computer data, repeating the reasoning above, before realizing that there is something crucially different between things amenable to theft and those that are not. The first notable case dates from 1983.[4] It concerns a programmer taking a disk pack[5] from his former employer and using the source code stored on the disks to develop a competing software application. Some things were completely clear, including that taking the disk pack qualifies as theft. But what about the data on the disk? Were these stolen? The court,

* Ronald Leenes is professor of regulation by technology at the Tilburg Institute for Law, Technology, and Society at Tilburg University, the Netherlands. I am grateful to comments on earlier drafts by Bert-Jaap Koops, Leonie Reins, Aaron Martin, Giorgio Monti, and Margot Hol. The usual disclaimers about the final result apply.

1 The Supreme Court adopts a restricted interpretation of assets and considers Intellectual property, such as copyright and patents, out of scope.
2 Also known as *Nulla poena sine lege previa*.
3 HR 11 mei 1982, NJ 1982/583, m.nt. 't H.
4 Hof Arnhem (strafkamer) 27 oktober 1983, NJ 1984, 80 CR 1984-1, p.31, m.nt. J.M.Smits, (Computergegevens).
5 https://en.wikipedia.org/wiki/Disk_pack.

to the dismay of some scholars, adopted the same reasoning as the Supreme Court had done in the 1921 and 1982 cases.

Then, in 1995/1996 two cases decided by the Supreme Court changed the course, settling that bits are not to be treated as atoms.[6] The 1995 'PIN code' case was a first eye-opener.[7] The case involved an assault in which the victim was deprived of his bank card (while drawing money from an ATM) and was forced to disclose the PIN code to the robber. The Supreme Court realized that in such a case the possessor of the PIN code does not lose it as an effect of disclosing it and that only a copy is provided, unlike the theft or extortion of a tangible good. The 'multiple' nature of computer data (more people can have possession of them at the same time) makes them fundamentally different from physical goods.

The same reasoning was followed in the Supreme Court's 'computer data' ruling,[8] in which a network manager had copied files without permission of the owner of the computer system (similar to the 1983 case mentioned above). The Supreme Court here moves back to the question whether computer data are 'goods' instead of approaching the issue along the lines of protectable assets. The Court re-iterates that for embezzlement (art. 321, Criminal Code) to be applicable, computer data should qualify as 'goods'. This is not the case with computer data according to the Court because this requires 'the holder to lose exclusive control over the data', which is not the case here. The system's owner can still access the original data, the network manager only had obtained a copy.

The Supreme Court in these two cases acknowledges that the traditional provisions for theft, extortion and similar in the Criminal Code do not cover acts involving making copies of intangibles.

It looked as if the legislator and courts had herewith definitively settled the matter – bits are not to be treated under the atoms-based provisions in criminal law – and thus addressed the foundational friction in the law caused by the rise of computer technology. However, in the 2010s, new developments in digital technologies reopened the struggles of courts with the properties of atoms and bits, the Runescape[9] and phone credit (Belminuten)[10] cases. In the Runescape case, a player was forced to hand over a virtual good (a mask and an amulet). The physical force took place in the real world but concerned virtual objects in the virtual world of the game Runescape. In this case, the various considerations raised in the earlier cases meet. The virtual objects are data (much like in the computer data cases), but there is exclusive control over the data (like in the cashless money case). The Supreme Court ruled that although the virtual objects are a type of computer data, they, like electricity share properties of assets worthy of protection against theft and extortion because they represent value and furthermore they exhibit exclusive use. Virtual goods can therefore be the object of theft in criminal law. A similar reasoning was adopted in the phone credit case, which dealt with a stolen SIM card that contained credit for making phone calls and sending text messages.[11] This made sense in the context of a digital environment in which some computer data constitute unique objects whose value can be used by only one person at the same time, rather than multiple objects whose value can benefit several people simultaneously. However, it also opens up a new area of uncertainty – and therewith new friction – since now, courts will have to assess whether computer data in a particular case are to be treated as similar to atoms (in the line of the electricity judgement) or as similar to bits (in the line of the 1996 computer data judgement).

What we see in the cases discussed is that the courts cope with new situations through, for instance, teleological interpretation and by expanding and contracting the scope of concepts. What is at play could be described as an attempt to maintain (dual) *coherence*[12] in law: "the reading that is adopted must maintain a thread of continuity with the jurisprudence; and, secondly, the reading must cohere with the constitutive (moral) values of a particular legal order".[13]

1.1 The new kid on the block

Law is a living, flexible system and has ways of accommodating new situations and phenomena. Sometimes the changes induced by new technologies are profound and have the potential to significantly disrupt the law. The emergence of Cyberspace was such a change. Although lawyers are said to be slow in picking up technological changes, it would be fair to say that the famous *Law of Cyberspace Conference* at the University of Chicago in 1996 was an example of legal scholars seeing early where the puck is heading. The conference made at least two people famous: Judge (and professor) Frank H. Easterbrook and professor Lawrence Lessig. It assembled a group of enthusiastic legal scholars who saw the legal challenges of the Internet coming and discussed the prospects of Cyberlaw, the law needed to regulate this new space. Easterbrook, however, immediately threw water on the enthused spirit in his keynote address called "Cyberspace and the Law of the Horse".[14] The passage of his keynote that drew most attention referred to a claim by former dean of the University of Chicago Law School, Gerhard Casper, that teaching the 'Law of the Horse' would be nonsense. With the Law of the Horse, he meant the legal body of knowledge relating to all things horses, including sales of horses, injuries caused by horses, licensing and races of horses etc.[15] Easterbrook extends this argument to Cyberspace. There is no need for specialized or niche legal studies applied to Cyberspace:

> "...the best way to learn the law applicable to specialized endeavors is to study general rules. Lots of cases deal with sales of horses; others deal with people kicked by horses; still more deal with the licensing and racing of horses, or with the care veterinarians give to horses, or with prizes at horse shows. Any effort to collect these strands into a course on 'The Law of the Horse' is doomed to be shallow and to miss unifying principles."[16]

6 The legislator had by then incorporated changes in the Criminal Code, based on the finding that computer data are not to be considered 'goods' under criminal law. For instance, article 317 Dutch Criminal Code (extortion) was amended to include, besides forcing someone to hand over a 'good', 'to make available information with monetary value in business and trade' ('het ter beschikking stellen van gegevens met geldswaarde in het handelsverkeer' in Dutch). Wet computercriminaliteit, Staatsblad 1993, 33. Since the events leading up to the Supreme Court cases took place prior to this, the legislative change did not affect the case.

7 HR 13 juni 1995, ECLI:NL:HR:1995:ZD0064.

8 HR 3 december 1996, LJN ZD0584, NJ 1997, 574 (Computergegevens).

9 Hoge Raad 31 januari 2012, LJN BQ9251.

10 Hoge Raad 31 januari 2012, LJN: BQ6575.

11 Hoge Raad 31 januari 2012, LJN: BQ6575.

12 Roger Brownsword, 'Regulatory Coherence—A European Challenge' in Kai Purnhagen and Peter Rott (eds), *Varieties of European Economic Law and Regulation: Essays in Honour of Hans Micklitz* (Springer 2014); Roger Brownsword, *Law, Technology and Society: Re-Imagining the Regulatory Environment* (Routledge 2019).

13 Brownsword, *Law, Technology and Society* (n 12) 134.

14 Later published as Frank H. Easterbrook, 'Cyberspace and the Law of the Horse' [1996] *University of Chicago Legal Forum*.

15 Note that this remark has to be placed in a Common Law context where the law primarily consists of case law.

16 Lawrence Lessig, 'The Law of the Horse: What Cyberlaw Might Teach' (1999) 113 *Harvard Law Review* 501, 502.

This claim has had a profound impact on the emerging field of cyberlaw and I would dare say the echoes of Easterbrook's remarks still resonate today.

Easterbrook's insistence on the value of general principles in teaching the law is understandable. These principles provide coherence[17] – integrity and internal consistency – in the law and make understanding what the law requires of us easier as well as provide legal certainty.[18] Scholars in the emerging field of cyberlaw were quick to respond. Lawrence Lessig, for one, tried to counter Easterbook's claim that focusing on law in cyberspace does not shed insights on unifying principles. He argued that looking at how cyberspace interacts with law we learn something about the "limits on law as a regulator and about the techniques for escaping those limits. [...] By working through these examples of law interacting with cyberspace, we will throw into relief a set of general questions about law's regulation outside of cyberspace".[19] In particular, he draws attention to the fact that cyberspace brings a new modality of regulation, "code", which in his words comprises the hardware and software that make up the Internet.[20] Code has turned out to be very powerful regulator indeed. Leaving aside that the regulative and normative effects of artifacts are nothing new, certainly not for philosophers of technology[21] and science and technology studies (STS) scholars, the message that code/architecture/design in fact regulates human behaviour and as such can be placed in line with law as a regulatory instrument, certainly was a new message for legal scholars.

Is this the kind of general lessons Easterbrook expected in order to count as being on par with 'tort' or 'contract'? No, certainly not. Andrew Murray and others are probably right that Lessig "failed to rebut key indictments in Easterbrook's challenge to the Cyberlaw community, [and that] instead he simply pled 'special circumstances'"[22]. And so the debate has continued and, in fact, this editorial marks just one step in it.

While the discussion alluded to above concerned cyberspace and the attempt to start getting our heads around regulating this novel space through cyberlaw, also other technologies have presented themselves or move from the realm of science fiction to everyday life. Biotechnology and especially genomics made great progress in the second half of the 1990s and entered the academic agenda around the turn of the millennium.[23] Around 2005, nanotechnology came to the fore in legal scholarship, challenging existing distinctions in law once again: should titanium dioxide particles in sunscreens be treated as cosmetics (not penetrating the skin) or drugs (which do)? Nanotechnology also raised new regulatory challenges,[24] including how to regulate nanocarbon (nanotubes, etc.), and whether the 'grey goo' scenario (nanorobots self-replicating to form an ever more consumptive grey goo[25]) called for regulatory intervention. Robotics, artificial intelligence, cloud computing, and blockchain followed suit.

Every time a new technology gains traction, the same questions are asked. What are the ethical and legal issues raised by the technology and how is it regulated in the first place? Many who have been engaged in this kind of quest have experienced the ghost of the Law of the Horse. Each time a new technology is put on the table, it feels like trying to fit the technology in the existing concepts, categories and classifications, while at the same time looking for the X-law.[26] And each time the conclusion seems to be that there is a patchwork of applicable traditional concepts (property, liability, privacy, etc.) that cover part of the issues surrounding the new kid on the block and apart from the generic doctrines there is a patchwork of specific legal frameworks that deal with other aspects. And of course, lacunae, inconsistencies, and undesirable effects are found as well. On occasion, the technologies defy being forced into the existing classifications on which coherence in law is built.

Is this friction with legal coherence specific to new technology or technologies in general? I do not think so. Coherence is (becoming?) an issue elsewhere as well. Society is becoming ever more complex and the traditional concepts and institutions increasingly become inadequate to deal with this complexity. As a case in point, civil law professor Stephanie van Gulijk in her inaugural address at the Tilburg Law School led the audience through the complex network of entities involved in construction and how no one legally is responsible for the safety of buildings (with the collapse of a parking garage at Eindhoven airport in 2017 as an example).[27] The existing legal framework is primarily aimed at bilateral arrangements and is repressive in nature[28] and has difficulties in coping with complex conglomerates of actors that deal with buildings involving novel concepts such as Design Build Finance Maintain & Operate (DBFMO), Design Build Maintain & Remove and DBFMO-Deconstruct, where the involvement of partners may well extend the initial construction phase.

2. Identity crisis

Of course, building requires technology and is technology, but it is not the kind of technology many of us[29] in the field of technology and law have in mind when discussing technology regulation.[30] Our interest is technology with a capital T, so let us return to our common interest. Cyberlaw and the Law of the Horse has been troubling schol-

17 See Roger Brownsword, 'Law Disrupted, Law Re-Imagined, Law Re-Invented' [2019] *Technology and Regulation 11, 17*; also see Michael Guihot, 'Coherence in Technology Law' (2019) 11 *Law, Innovation and Technology* (forthcoming).

18 See also Arthur Cockfield and Jason Pridmore, 'A Synthetic Theory of Law and Technology' (2007) 8 *Minnesota Journal of Law, Science & Technology* 39, 496.

19 Lessig (n 14) 502.

20 Lessig (n 14) 506.

21 Langdon Winner, 'Do Artifacts Have Politics?' (1980) 109 *Daedalus* 121. In 1977 already, Winner wrote "[...] the crucial awareness that technology in a true sense is legislation. It recognizes that technical forms do, to a large extent, shape the basic pattern and content of human activity in our time. Thus politics becomes (among other things) an active encounter with the specific forms and processes contained in technology." (italics in the original) Winner, L. 1977. *Of Autonomous Technology: Technics-out-of-Control as a Theme in Political Thought*: MIT Press, 232.

22 Andrew Murray, 'Looking Back at the Law of the Horse: Why Cyberlaw and the Rule of Law Are Important' (2013) 10 *Scripted* 311.

23 E.g., Roger Brownsword, WR Cornish and Margaret Llewelyn (eds), *Human Genetics and the Law: Regulating a Revolution* (Hart 1998).

24 EJ Koops and others, 'Een heel klein artikel met grote gevolgen. Eerste verkenning van nanotechnologie & recht' (2005) 80 *Nederlands Juristenblad* 1554; Bert-Jaap Koops and others, 'On Small Particles and Old Articles - An Exploration of Legal and Regulatory Issues of Nanotechnologies' (Social Science Research Network 2008) SSRN Scholarly Paper ID 1300925 https://papers.ssrn.com/abstract=1300925 accessed 17 April 2019.

25 Eric Drexler, *Engines of Creation* (Anchor Press/Doubleday, 1990); Michael Crichton, *Prey* (Harper 2002).

26 We have done so in the Robolaw project (http://robolaw.eu; Ronald Leenes and others, 'Regulatory Challenges of Robotics: Some Guidelines for Addressing Legal and Ethical Issues' (2017) 9 *Law, Innovation and Technology* 1), but have seen similar impulses in other EU and national projects.

27 Stéphanie van Gulijk, *Circulair en veilig bouwen. Verantwoordelijkheid is geen estafettestokje* (Tilburg University 2019).

28 Gulijk (n 27) 28.

29 I will refer to us as the legal scholars interested in technology regulation and associated fields, but maybe the scope of 'us' is much wider, as we will see.

30 Unless it concerns Smart Homes and Smart Buildings.

ars in 'our' field over the years[31].

On a possible bright side, as Brownsword rightly notes, Easterbrook was wrong in his prediction that cyberlaw would have no future. *Technology law*, and regulation of and by technology, has become a distinct area of scholarship, has research institutes devoted to its study[32], has its own journals[33] and has taken solid ground in teaching as well[34].

The fact that the journal *Law, Innovation and Technology* still publishes papers that refer to the Law of the Horse[35] signals that we are not done yet, or slightly more negatively, that 'we' still suffer an identity crisis. The symptoms of this crisis relate to (in increasing order of severity) our posing of the same questions:

- Is there such a thing as technology law?
- What are the boundaries and scope of 'our' field?
- Who are 'we'?
- How to regulate technology?
- What might Cyberlaw/Robolaw/Ledgerlaw teach?
- What is the role of law in a world that increasingly is driven by technologically spurred innovation?

In the following, I will try to sketch the outlines of the field that I would designate technology regulation and introduce this journal as a means to further this field.

2.1 Regulation

There is a large body of scholarship on all these topics. For instance, Lyria Bennett Moses[36] has argued that technology is not particularly suited as a regulatory target and that technology regulation is the wrong designation of the field. Besides the fact that the term regulation triggers different meanings with different people and is potentially both broader and narrower than law, 'it is generally not the technology that is regulated, but rather a socio-technical landscape'. She is right in this, but for the moment I will maintain technology regulation as a convenient shorthand.

Regulation indeed is a problematic concept. As Karen Yeung has observed, the meaning of regulation is notoriously inexact and highly contested.[37] Within the realm of technology regulation, however, there seems to be agreement that regulation affects the behaviour of individuals and (often) restricts their autonomy and freedom to act. Within this frame, regulation hence is much broader than just command and control rules enacted by the state. Instead, and moving away from the state as sole regulator, a relatively established definition of regulation is Julia Black's: 'Regulation is the sustained and focused attempt to alter the behaviour of others to standards or goals with the intention of producing a broadly identified outcome or outcomes, which may involve mechanisms of standard-setting, information gathering and behaviour-modification.'[38] This clearly brings into scope what Lessig has framed as 'code', architecture or design[39] or what has been commonly known as techno-regulation[40], the "deliberate employment of technology to regulate human behaviour"[41], or as Koops[42] formulates it: "technology with intentionally built-in mechanisms to influence people's behaviour". Markets and social norms also fall within Black's regulatory framework.

Much has been written about techno-regulation, including that there is a whole spectrum of technology-mediated forms of behavioural influence. But there are still many questions regarding the nature and scope of techno-regulation, for instance, is *intent* a necessary component of behavioural modification or do side-effects of design (a CD player cannot play DVDs, although the disks look the same) also count as behavioural modification? Is a wall-socket techno-regulation? If so, what does it regulate? Wall-sockets and plugs do limit my ability to use appliances abroad, but is that regulation as we mean to discuss it? Other characteristics of the spectrum of techno-regulation are also not entirely understood.[43] Techno-regulation incorporates family members as varied as *affordances*[44], *nudges*[45], *persuasive technologies*[46], instrumental techno-regulation enforcing existing legal norms (for instance a speed bump) and intrinsic techno-regulation constituting the norm itself (design choices that limit certain uses of technology) ranking differently on aspects such as (user) choice, (user) awareness and compulsion.[47]

2.2 Technology

The scholarship on (techno-)regulation does not resolve the boundary issue of the field of technology regulation. Although Bennett Moses shows that we should not focus on technology as regulatory targets, but rather at socio-technical systems, that insight only leads us somewhat along the way. Many technology regulation scholars seem tempted to focus on new, emerging or disruptive technologies and the (novel) issues these raise. This provides the gratification of being at the forefront of development and not be bogged down with 'old'

31 Starting perhaps with Lessig (n 16) 502, but in general, this is what unites much of the works cited in this editorial.

32 Such as my academic home, the Tilburg Institute for Law, Technology, and Society (TILT), which has been around since 1994.

33 Such as Law, Innovation and Technology (LIT).

34 Such as the MA program in Law & Technology run by TILT.

35 E.g., Guihot (n 17).

36 Lyria Bennett Moses, 'How to Think About Law, Regulation and Technology: Problems with "Technology" as a Regulatory Target' (2013) 5 *Law, Innovation and Technology* 1.

37 Karen Yeung, 'Towards an Understanding of Regulation by Design' in Roger Brownsword and Karen Yeung (eds), *Regulating Technologies: Legal Futures, Regulatory Frames and Technological Fixes* (Hart 2008), 90. To some, regulation refers to 'command and control'; rules enacted by government (top-down), enforced by sanctions (e.g., Ian Ayres and John Braithwaite, *Responsive Regulation: Transcending the Deregulation Debate* (Oxford University Press 1992)). Some scholars restrict the scope of regulation to state intervention, while others include any actor or institution that can regulate human behaviour within the scope of regulation. According to the former, measures introduced by the market, such as the region codes in DVD players (an example of techno-regulation or regulation by design) are, by definition, not forms of regulation, whereas it constitutes regulation according to the latter perspective.

38 Julia Black, 'Critical Reflections on Regulation' (2002) 27 *Australian Journal of Legal Philosophy* 25.

39 Andrew Murray and Colin Scott, 'Controlling the New Media: Hybrid Responses to New Forms of Power' (2002) 65 *Modern Law Review* 491.

40 The term techno-regulation was, as far as I am aware, introduced by Roger Brownsword. Roger Brownsword, 'What the World Needs Now: Techno-Regulation, Human Rights and Human Dignity' in Roger Brownsword (ed), *Global Governance and the Quest for Justice* (Hart Publishing 2004).

41 Ronald Leenes, 'Framing Techno-Regulation: An Exploration of State and Non-State Regulation by Technology' (2011) 5 *Legisprudence* 143; RE Leenes, *Harde lessen: Apologie van technologie als reguleringsinstrument* (Tilburg University 2010).

42 Bert-Jaap Koops, 'Criteria for Normative Technology. An essay on the acceptability of "code as law" in light of democratic and constitutional values' in *Regulating Technologies* (Roger Brownsword and Karen Yeung, eds.), Oxford: Hart Publishing 2008, 158.

43 Bibi van den Berg and Ronald E Leenes, 'Abort, Retry, Fail: Scoping Techno-Regulation and Other Techno-Effects' in M Hildebrandt and AMP Gaakeer (eds), *Human law and computer law* (Springer 2013).

44 Donald A Norman, *The Psychology of Everyday Things* (Basic Books 1988).

45 Richard H. Thaler and Cass R. Sunstein, *Nudge: Improving Decisions about Health, Wealth, and Happiness* (Yale University Press 2008).

46 BJ Fogg, *Persuasive Technology: Using Computers to Change What We Think and Do* (Morgan Kaufmann Publishers 2003).

47 van den Berg and Leenes (n 43).

technologies. But the truth is that we seem fairly arbitrary in picking our targets of attention. In fact, technology is a problematic term in itself given its wide scope. Looking at dictionaries and scholarly works we see definitions such as 'technologies comprise the broad range of tools and crafts that people use to change or adapt to their environment'.[48] That makes paper a prominent technology[49], but is it one worth discussing in LIT or this journal? Are we, or should we be interested in discussions about regulating paper or the use of paper? This is a relevant question in determining the scope of the field.

Of course there is regulation regarding paper, for instance regarding the production or disposal of paper, but that seems more the realm of environmental law than of technology law/regulation. Moving a little away from this, we enter the realm of publishing and freedom of expression/speech. With that we enter media law. Which parts of media law and freedom of expression are part of the domain of technology regulation and which are not?

Also, all technologies lose their novelty sooner or later. At what point are they no longer of interest to us? An intuitive, or maybe tautological, answer would be, when they no longer raise legal disruptions or significant legal frictions. However, technologies tend to develop, or rather are being developed by humans, and acquire new features and functions. Hence it is not the case that a technology on a larger scale (e.g., paper) per se is out of interest for technology scholars, but rather new applications or uses require or draw attention. Rarely do we encounter entirely new classes of technologies. Steam engines, computers, and the Internet surely are major new technologies, but arguably most social media, for instance are novel incarnations of discussion fora of old.

Another question is what the appropriate scale of a technology is to merit our attention. Paper could be the target of choice, but so do political speech in writing, or advertisements in magazines. All levels can be studied and regulated. Whether they do will largely depend on the legal frictions induced by the use of technology.

2.3 The 'we' in technology regulation

As part of any proper identity crisis, reflections on what we are, and why, are inevitable. Andrew Murray gave a wonderful keynote address[50] at the 2013 Bileta[51] conference that bears witness to precisely this point. The backdrop of his presentation is much like the present editorial, looking back at 'The Law of the Horse' and what 'regulatory cyberlawyers'[52] have to offer to define an agenda for the future.

In his struggle with rebutting Easterbrook, Murray makes a number of observations that should sound familiar to many who consider themselves in the genus 'regulatory technology lawyer' (techlawyers for short) as I, for the time being, would want to call the legal scholars working in the field of technology regulation.

Cyberregulatory or cybergovernance theorists are convinced that digitisation and cyberspace are special (more on exceptionalism below). Murray explains how he and others in their analysis employ 'academic heavyweights – Michael Foucault, Bruno Latour, Niklas Luhmann – and a number of legal academic cruiserweights – Gunther Teubner, Cass Sunstein, Neil MacCormick – to make our point that Cyberspace and cyber-regulation is special. The problem is we continue to use the language and rhetoric of social policy, sociology and political philosophy, rather than the language of law or regulation.'[53] To then conclude '[w]e become social scientists not lawyers', and as a colleague told Murray, 'what you do isn't law'[54].

These observations do resonate with me at least. But I think the observation that cyber/techlawyers move beyond the law is precisely the point of what they do. They acknowledge that cybergovernance and technology regulation require multidisciplinarity and that its scholars should be versed in more than just the law. Black letter law is less essential in technology regulation than in more traditional legal fields because there is more to regulation than law and because technology has the potential to disrupt classical legal concepts and institutions and sometimes does. Understanding 'the interplay between law and technology and the ways technology can have a substantive impact on individuals and their legal interests apart from the technology's initial intended use'[55] becomes essential in this respect.

Hence, it should not come as a surprise that the field of technology regulation is populated by others than legal scholars. And as is the case in many realms within academia, the field is heavily balkanized. Already mentioned are philosophers of technology and STS scholars as members of the broad family of technology explorers. They have their own (respective!) perspectives and methodologies, but are generally interested in the relation between technology and moral concepts.

Next to the philosophical branch of the family, there is also an economic branch. Anna Butenko and Pierre Larouche[56] have pointed out that in the legal literature at the interplay between innovation and law, there are two related fields of study that are not commonly brought together. One is law and economics as it concerns innovation, which is largely examining the effects of innovation, and the mechanisms to stimulate innovation in a market economy. The second is law and technology, which conflates largely with the area we have been discussing above, and which, according to Butenko and Larouche, often investigates either technology as a regulatory focus and rationale for regulating, or regulation by technological means. Both fields deal with the regulation of innovation, but are usually separate disciplines.[57] I see both fields of study as part of the wider field of technology regulation that I am tentatively framing in this editorial.

The question of who else belongs to the family of technology regulation or who else we need to build a coherent theory of technology regulation is an open question. I hope this journal will contribute to answering this question.[58]

48 Bert-Jaap Koops, 'Ten Dimensions of Technology Regulation. Finding Your Bearings in the Research Space of an Emerging Discipline' in Morag Goodwin, Bert-Jaap Koops and Ronald Leenes (eds), *Dimensions of Technology Regulation* (Wolf Legal Publishers 2010).

49 And indeed Mireille Hildebrandt has written interesting works about paper technology, for instance in relation to law. See for instance, Mireille Hildebrandt, 'Technology and the End of Law' in Erik Claes, Wouter Devroe and Bert Keirsblick (eds), *Facing the Limits of the Law* (Springer 2009).

50 Murray (n 12).

51 The British and Irish Law, Education and Technology Association

52 Murray (n 22) @314 distinguishes between regulatory cyberlawyers (like himself) and 'applied cyberlawyers' while admitting that certainly the latter term is not ideal.

53 Murray (n 22) 314.

54 Murray (n 22) 314.

55 Cockfield and Pridmore (n 18) 503.

56 Anna Butenko and Pierre Larouche, 'Regulation for Innovativeness or Regulation of Innovation?' (2015) 7 *Law, Innovation and Technology* 52.

57 Not so at The Tilburg Law School, where, TILEC (law and economcs) and TILT (law and technology) have been united in the new department of LTMS.

58 One of the panels at the *TILTing* 2019 conference was devoted to questions such as these, and likely we will hear from the panelists (Michael Guihot, Lyria Bennett Moses, Roger Brownsword, Bert-Jaap Koops, Han Somsen, Ronald Leenes) sometime soon.

2.4 What technology regulation might teach

In discussions with others and even to earn our spot under the sun, discussions about the boundaries of the field are relevant, but the subject of our field is more so.

The central concerns could be phrased as: what (new) issues are or could be created due to technology development and if so, how should we regulate this technology (instance/use)?

All too often, people, scholars, policy makers, industry and interest groups, jump to the conclusion that regulation is required, often conveniently accompanied with recommendations as to what that regulation could or should look like, opening the door widely to regulatory capture. The 'flawed law syndrome' is very prevalent in technology circles.[59] Regulating before understanding what is at stake (the particular technology), what the issues are, for whom, why and what is wrong or missing in existing regulation, if anything, is not the proper starting point. But how to systematically go through the steps and questions that do make sense is difficult without proper methodologies and frameworks. We do have some frameworks or theories that warrant further reflection and development such as Arthur Cockfield and Jason Pridmore's 'Synthetic Theory of Law and Technology',[60] which will be briefly discussed below, and Roger Brownsword's 'Re-invention of Law' in view of the technological disruption of law and legal reasoning[61].

Supposing that we have answers to the non-trivial questions regarding issues, stakeholders, values, etc., the questions become whether, when and how to regulate. Here we see much scholarship and also clear (implicit) differences between scholars and their cultures. From a continental European perspective, regulation enacted by the EU or national legislators is a legitimate starting point. We live in an area with a regulatory-instrumentalist mindset as Brownsword calls this approach to regulation.[62] There generally are regulatory purposes and policies following from public interest, social justice, or market failure that call for regulation and guide its direction. Coming from a law and economics perspective, or from the US regulatory mindset, this approach to governing society is less obvious. In these realms, addressing market failure is a legitimate reason to interfere through regulation; other reasons of public interest (who defines these?) far less so.

Regarding the timing of regulation, we have clearly learned lessons. The law is said to always lag behind technological development and again significant scholarship exists here.[63] The pacing problem or regulatory connection[64] is well known and so is what has become known as the Collingridge dilemma —"When change is easy, the need for it cannot be foreseen; when the need for change is apparent, change has become expensive, difficult, and time-consuming."[65]

There is also reflection and scholarship on policy heuristics (i.e. one-liners) devised during the late 1990s as a way to guide legislators in coping with the Internet and other ICT developments.[66] Included were classic notions such as, 'what holds offline, should hold online as well' and regulate through technology-neutral regulation. Some of these heuristics were clearly based on maintaining a congruence between rules in the real world and the rules in cyberspace, which for the sake of legal certainty should be recognizable for cybernauts. This idea had its flaws at the beginning of the Millennium already, but one may certainly wonder whether this congruence is maintainable and desirable in 2019. It presupposes an off-line experience prior to entering cyberspace. Current teenagers lack this pre-cyberspace experience and do not so much have to make the move from atoms to bits.[67] They, for instance, have hardly have experienced music and other content in forms sold in brick-and-mortar shops. The excludability and rivalry characteristics of physical carriers protected by copyright are almost alien to people born digital for whom that song is always just one click away.

2.5 Is technology regulation destined to lead to bad law?

Chris Reed, in discussing the substance or way technology is regulated notes that technology regulation moves in particular directions, leading to 'bad law'.[68] He argues that "[T]here is a clear trend for law and regulation, particularly in cyberspace, to become increasingly precisely specified. The perceived benefit of this approach, increased certainty as to compliance, may be illusory. Over-complex laws have serious disadvantages, particularly a greatly weakened normative effect, and problems of contradiction and too-frequent amendment."[69] Although this seems plausible enough as an argument, I am not convinced by his explanation nor by the examples he gives, but that is for another occasion.

Reed does rightly point at a bigger underlying problem, regulatory disconnect and its cousin regulatory failure. He seems to suggest that technology regulation, in part due to wrong choices by the regulator on the dimensions 'vagueness-certainty', 'opaqueness-clarity', and 'complexity-simplicity', is almost destined to lead to regulatory failure. One of the problems here is that the notion of regulatory failure is underdeveloped. Failing, but compared to what? Policy goals, for instance. But what if these are unclear. In a study of one of the cases that could qualify as regulatory failure, the European cookie regulation, analysis of the Dutch policy and legislative debate reveals that there is no political consensus regarding the policy goals.[70] In view of this disagreement, the regulation maybe does what it is supposed to do given unclear goals.

Reed raises a number of relevant questions that warrant further explo-

59 Ronald Leenes, 'Regulating New Technologies in Times of Change' in L. Reins (ed), *Regulating New Technologies in Uncertain Times* (TMC Asser 2019).
60 Cockfield and Pridmore (n 18).
61 Brownsword (n 17)
62 Brownsword (n 17) 15.
63 E.g., Lyria Bennett Moses, 'AGENTS OF CHANGE: How the Law Copes with Technological Change' (2011) 20 *Griffith Law Review*, Vol. 20, No. 4, 764; Roger Brownsword and Morag Goodwin, *Law and the Technologies of the Twenty-First Century: Text and Materials* (Cambridge University Press 2012).
64 E.g., in Brownsword and Goodwin (n 65); Diana Bowman, 'The Hare and the Tortoise: An Australian Perspective on Regulating New Technologies and Their Products and Processes' [2013] *Innovative Governance Models for Emerging Technologies* 155; Roger Brownsword and Han Somsen, '"Before We Fast Forward – A Forum for Debate"' (2009) 1 *Law, Innovation and Technology* 1.
65 The quote comes from Evgeny Morozov, 'The Collingridge Dilemma' in J Brockman (ed), *This explains everything* (Harper Perennial 2013). The original concept is discussed in David Collingridge, *The Social Control of Technology* (Frances Pinter 1980).
66 Bert-Jaap Koops and others (eds), Starting Points for ICT Regulation: Deconstructing p[r]Evalent Policy One-Liners (TMC Asser 2006).
67 Nicholas Negroponte, *Being Digital* (Vintage Books 1996).
68 Chris Reed, 'How to Make Bad Law: Lessons from Cyberspace' (2010) 73 *The Modern Law Review* 903; Chris Reed, *Making Laws for Cyberspace* (Oxford University Press 2012).
69 Reed, 'How to Make Bad Law: Lessons from Cyberspace' (n 56) 903.
70 Ronald Leenes, 'The Cookiewars: From Regulatory Failure to User Empowerment?' in Marc van Lieshout and Jaap-Henk Hoepman (eds), *The Privacy & Identity Lab* (The Privacy & Identity Lab 2015).

ration. The rules/standards/principles debate touched upon in his work clearly is a centerpiece, but not only for cyberlaw. The amount of judgment left to regulatees is a fundamental question that relates to comprehensibility, 'compliability', etc. But we should also take into account that not all regulatees are alike. Kagan and Scholz[71] provide some guidance in this respect. They distinguish *amoral calculators*, who make cost-benefit assessments and then determine whether to comply with the rules or not. The content of the rules does not matter, the fines do. A different group is that of the *political citizens* who do not follow certain rules as a matter of civil disobedience. And then there are the *organisationally incompetent*. These are the ignorati, they do not know or understand the rules. We need to be aware that all three types operate in the same space and we should not assume too easily that the rules are inadequate.

As part of the regulatory toolbox that goes beyond traditional law, 'smart regulation' or 'responsive regulation'[72] and 'participatory governance',[73] should be mentioned. They may contribute to the regulatory innovation[74] necessary to address the regulatory challenges of complex technological developments that have broad and systemic implications for many social processes.

3. Back to the Horse

I started my recount of the field of cyberlaw with 'The Law of the Horse'. In the meantime other animals have joined the herd as metaphors for the field of technology regulation or subfields thereof.

Michael Guihot[75] attempts to outline the boundaries of the broader domain by defining technology law as a relatively coherent field, similar to environmental law or health law, shedding light on the complex interaction of participants, pressures and regulatory responses in view of technology development. His paper contains an image where technology law sits in the middle of five core legal frameworks: contract, property, privacy, tort, and competition law. The image shows Technology Law as the face of a fox in-between the circles that depict these five fields. Guihot pledges the field to be called *technology law* instead of *technology regulation* if it seeks to be included in the canon of legal fields. If that is the aim then he may have a point.

I do, however, think that inclusion in the canon is not the sole ambition of the field. In my view the field is not only a body of legal knowledge, but also a field that studies how to regulate technology or socio-technical assemblages. It not only provides guidance on *what is*, but also about *what might be*. That may warrant seeing technology law in the form of the codified legal knowledge (statutes and case law) as a potential subfield of technology regulation next to theory and methodologies suitable for regulators.

Coming from a background in mathematics, Bert-Jaap Koops has provided a starting point to define what he then termed the relatively new field of technology regulation by spanning it up in ten dimensions.[76] As he remarks, most people have great difficulties in imagining anything beyond three or four dimensions and hence comprehending what exactly the space is spanned by the ten dimensions, his model at least allows people to 'see where you are, or where you want to go, in the technology regulation space, all you have to do is determine the coordinates along ten different dimensions.'[77] The model can be summarized by noting that it entails three regions: regulation, with dimensions of knowledge, normative outlook, and type of regulation; the technology region, with technology type, innovation, place, and time as subdimensions; and research region, spanning discipline, problem and frame. This brief overview already shows that many different types of technologies, modes of regulation and types of research can find home in this framing of technology regulation.

Arthur Cockfield and Jason Pridmore have outlined a synthetic theory of law and technology that can inform law and technology analysis.[78] They want to move away from a 'traditional compartmentalized approach that scrutinizes niche doctrinal areas of technology law (e.g., patent law or copyright law) or the impact of specific technologies (e.g., cyberlaw, new media, or biotechnology)'[79] and instead look at the broader implications of technology on law. Their theory prescribes two steps. In the first it needs to be established whether technological change undermines traditional interests by identifying the traditional interest protected by law employing traditional doctrinal legal analysis and determining whether the interest is being duly disrupted by technological change. If technological disruption indeed is the case, a more contextual analysis is required. This analysis scrutinizes the broader context of technology change and its potentially unanticipated adverse outcomes for the traditional interest as well as for other protected interests the law seeks to protect. It then seeks to find legal solutions to protect the traditional interest that are less deferential to precedent and traditional doctrine.[80] This framework places the analysis of the intersection of law and technology squarely within a value/interest context. Instead of trying to fit in technologies within existing legal concepts and frames, it calls for taking a step back and re-evaluating underlying values to then determine new balances of interests and regulatory interventions to achieve these. At first glance this may resemble teleological reasoning as we have seen in the Dutch cases at the start of this paper; it is important to note that Cockfield and Pridmore call for a study of the technology in question and its further consequences for individuals and groups. This goes beyond classical teleological interpretation by courts.

Besides grand perspectives on technology law or technology regulation, there are also efforts to define subspaces. Han Somsen, for instance, has pointed at the inadequacy of environmental law in dealing with radical technologies and calls for a new regulatory effort and maybe even subfield.[81] Environmental law in his view aims more at environmental improvement (facilitating the cleanup of polluted rivers allowing salmon to return) than at environmental enhancement (genetically modifying salmon to cope with warmer waters due to global warming). 'Environmental enhancement, then, is an intentional technological intervention in the environment in pursuit of human interests, needs or rights which takes place outside the confines of such pre-agreed environmental base-lines.'[82] A base-line of environmental law is 'yes, unless', which may not be adequate to cope with radical climate engineering efforts (like colouring the ocean white to help lower global temperatures). He suggests we need reg-

71 R Kagan and J Scholtz, 'The Criminology of the Corporation and Regulatory Enforcement Strategies' in J Hawkins and J Thomas (eds), *Enforcing Regulation* (Kluwer 1984).
72 Ian Ayres and John Braithwaite, *Responsive Regulation: Transcending the Deregulation Debate* (Oxford University Press 1992).
73 Archon Fung and Erik Olin Wright, 'Deepening Democracy: Innovations in Empowered Participatory Governance' (2001) 29 *Politics & Society* 5.
74 Julia Black (ed), *Regulatory Innovation: A Comparative Analysis* (Elgar 2005).
75 Guihot (n 17).
76 Koops (n 48).
77 Koops (n 48) 312.
78 Cockfield and Pridmore (n 18).
79 Cockfield and Pridmore (n 18) 512.
80 Cockfield and Pridmore (n 18) 505.
81 Han Somsen, 'Towards a Law of the Mammoth? Climate Engineering in Contemporary EU Environmental Law' (2016) 7 *European Journal of Risk Regulation* 11.
82 Somsen (n 81) 119.

ulation on novel insights in values and capabilities of technologies. Although the scope and outlines of such regulation are left in the dark, Somsen has come up with a catchy name for this novel branch to environmental and technology law, *The Law of the Mammoth*.

Another notable effort to delineate a relevant technology subfield comes from Ryan Calo.[83] He places a law of robotics[84] next to cyberlaw as species of technology law. He argues that cyberlaw warrants being seen as a separate field and thus escaping Easterbrook's 'curse' because its 'introduction into the mainstream require[d] a systematic change to the law or legal institutions in order to reproduce or if necessary displace, an existing balance of values.'[85] Although robots share many qualities of the products of Cyberspace, embodiment, emergence and social valence makes them different with profound impact on 'a wide variety of contexts: criminal law and procedure, tort, intellectual property, speech, privacy, contract, tax, and maritime law, to name but a few'.[86] He goes on to show how frictions surface and concludes that robotics warrants an exceptionalist treatment in its own body of law. We see the urgency of changes along these lines around us. For instance, '[I]n the resolution from February 2019 on a comprehensive European industrial policy on artificial intelligence and robotics, the Parliament upheld its somewhat more proactive position on adopting new regulation in the field – it welcomed the Commission's initiative to create the Expert Group on Liability and New Technologies, but "regretted that no legislative proposal was put forward during this legislature, thereby delaying the update of the liability rules at EU level and threatening the legal certainty across the EU in this area for both traders and consumers'.[87]

There are likely many more efforts out there, and I think it is safe to say we do not have clearly established frames for 'the' domain or its subdomains.

4. The road ahead

In the brief overview I hope to have shown that the cyberhorse and other animals of cyberspace have not died (yet) and that many basic questions in the field of technology regulation are still not adequately answered. On the contrary. When a new technology presents itself, we witness new incarnations of the Law of the Horse. Amongst others, we see this regarding robotics, genomics and AI. Legislators and policy makers want to know what these phenomena legally are and whether or not (specific) regulation is required in coping with the changes the technologies induce. A first reflex then is to look for the law regulating this specific technology. What we find is that parts of the legal issues are covered by existing law (contract, tort), partly there may be specific provisions in these domains (bolted on existing concepts), partly there may be specific regulation, and there likely are undesired effects and regulatory gaps. Disruptive technologies are likely to lead to regulatory disconnect.

We need a more thorough theoretical, methodological and practical foundation to get a proper grip on technology and regulation. There is an urgency in doing so because the stakes are high, for instance because power is being concentrated in the hands of a limited number of (US and in the near future Chinese) players, and technology is being developed at a rapid pace. Some of the technologies have the potential to become Socially Disruptive Technologies (SDTs).[88] These technologies 'transform everyday life, social institutions, cultural practices and the organisation of the economy, business and work'.[89] Historical examples include the printing press, the steam engine, electric lighting, the computer, and the Internet.[90] Current candidates to receive the title of SDT include robotics, (general) Artificial Intelligence, gene editing, neurotechnology, and climate engineering. These examples seem to be technologies, but are in fact socio-technical systems. They have developers, creators, producers, users, affected non-users, constraints, requirements, consequences, etc. In most cases, there are many stakeholders involved and the resulting assemblages cross all sorts of boundaries, geographical (the machinery powering an AI in a car on the Dutch road may actually reside in the US, or more likely, somewhere in the Cloud) and hence jurisdictional, doctrinal, disciplinary, and so on.

The traditional coherence-based legal processes have difficulty keeping up with the changes induced by innovation and technology development. As Brownsword formulates it "coherentism presupposes a world of, at most, leisurely change. It belongs to the age of the horse, not to the age of autonomous vehicles"[91]. I think the field of technology regulation as broadly outlined above should strive to do better. Scholars in this young field do acknowledge the interactions between technologies, risks, and their regulation[92], or the interplay between regulation, technology and normative notions and values[93].

With the launch of the *Technology and Regulation* journal, the editors hope to offer a place to move the field forward. But why do we need a new journal for that, you may wonder?

4.1 Information wants to be free

Some of us are old enough to remember the pre or early Internet days. I clearly remember the telnet connections I had with colleagues in the US and how excited I was when I got Gopher running to browse the infosphere only just before Tim Berners-Lee gave us the World Wide Web. The mantra in those days was 'information wants to be free' and the development of the Internet and the Web took shape in this spirit. Search engines started appearing, making finding information scattered over the web easier and placing information more and more at our fingertips. Google's original 1998 mission statement was 'to organize the world's information and make it universally accessible and useful'.[94] Free, grass-roots initiatives further provided valuable information that has changed the world. Think of the Internet Movie Database (IMDb)[95], Wikipedia[96] and numerous general and specific information sources that many of us consult on a daily basis. Many of them are free of charge to users and maintained by donations and/or advertising.

In the meantime scholarly work to a large extent is not available to everyone free of charge, also not in the Internet sense of free (paid for by advertisements). Many journals are owned and run by commercial publishers that charge fees for their services. Publishers offer many

83 Ryan Calo, 'Robotics and the Lessons of Cyberlaw' (2015) 103 *California Law Review* 513.
84 Too bad no animals were introduced in this effort.
85 Calo (n 83) 552.
86 Calo (n 83) 553.
87 Rowena Rodrigues, '*Sienna D4.2: Analysis of the Legal and Human Rights Requirements for AI and Robotics in and Outside the EU*" (2019).
88 Philip Brey, 'Ethics of Socially Disruptive Technologies'.
89 Brey (n 72).
90 Brey (n 72).
91 Brownsword, (n 17) 10.
92 Guihot (n 17).
93 Cockfield and Pridmore (n 18).
94 https://www.theguardian.com/technology/2014/nov/03/larry-page-google-dont-be-evil-sergey-brin last consulted 12 May 2019.
95 IMDb's history predates the Web as a list on Usenet. It moved to the web in 1993. IMDb currently is owned by Amazon.
96 Launched only on January 15, 2001 https://en.wikipedia.org/wiki/Wikipedia, last consulted 12 May 2019.

useful services, such as facilitating quality control, offering reputation and brand, typesetting, distribution, storage and archival, monitoring and notification, etc. But this comes at a price. These costs are either borne by readers (subscription fees or one-off charges to access an article) or authors (article processing fees, etc.). And while this is not problematic for many scholars and other interested parties, this is certainly not the case for everyone in academia and beyond.[97]

The traditional subscription-based model is problematic due to declining budgets at universities. Choices have to be made, by libraries, departments and individuals, hardly anyone can maintain access to all relevant sources. For instance, Tilburg University, which lacks science departments, does not have subscriptions to ACM and IEEE journals. Yet, I have had the need to access these journals for my work in the European Commission's FP6/FP7/H2020 projects I have been engaged in. Of course there are workarounds to this issue, but I have seen the effects of limited access. And yet, Tilburg University is a relatively wealthy university. How do less fortunate researchers cope with these costs?

The Open Access (OA) funding models that aim to replace or complement the traditional model also have their issues. Researchers can include items in their budgets for Open Access Gold publication charges when applying for national or European grants (such as European Research Council grants). But many researchers do not have such projects with specific budgets for OA publication. This leads to tough choices within departments and schools, producing new have and have-nots.

This new journal, *Technology and Regulation*, offers an alternative. The costs of running the journal will be borne by the Department of Law, Technology, Markets, and Society (LTMS) embedded within the Tilburg Law School, facilitated by a grant from Tilburg University. We believe LTMS has the mass to perform the many tasks involved in running a professional, high-quality, peer-reviewed journal at zero costs for readers and authors.[98]

4.2 Where the rubber meets the road

Technology and Regulation is an international journal of law, technology and society, with an interdisciplinary identity. It will disseminate original research on the legal and regulatory challenges posed by existing and emerging technologies (and their applications) including, but by no means limited to, the Internet and digital technology, artificial intelligence and machine learning, robotics, neurotechnology, nanotechnology, biotechnology, energy and climate change technology, and health and food technology. As discussed above, regulation is conceived broadly to encompass ways of dealing with, ordering and understanding technologies and their consequences, such as through legal regulation, competition, social norms and standards, and technology design (or in Lessig's terms: law, market, norms and architecture).

Technology and Regulation aims to address critical and sometimes controversial questions such as:

- How do new technologies shape society both positively and negatively?
- Should technology development be steered towards societal goals, and if so, which goals and how?
- What are the benefits and dangers of regulating human behaviour through technology?
- What is the most appropriate response to technological innovation, in general or in particular cases?

It is in this sense that *Technology and Regulation* is intrinsically interdisciplinary: it is premised on the understanding that legal and regulatory debates on technology are inextricable from societal, political and economic concerns, and that therefore technology regulation requires a multidisciplinary, integrated approach. Through a combination of monodisciplinary, multidisciplinary and interdisciplinary articles, the journal aims to contribute to an integrated vision of law, technology and society.

Technology and Regulation invites original, well-researched and methodologically rigorous submissions from academics and practitioners, including policy-makers, on a wide range of research areas such as privacy and data protection, security, surveillance, cybercrime, intellectual property, innovation, competition, governance, risk, ethics, media and data studies, and others.

The journal opens with this editorial and two invited papers. Regular papers in Technology and Regulation are double-blind peer-reviewed and completely open access for both authors and readers. It does not charge article processing fees. Technology and Regulation is an online journal with rolling publication. The journal publishes papers as fast as the editorial team and reviewers can process them. The published papers are available as self-contained PDFs with all the relevant elements, such as page numbers, DOI, ISSN, etc.

Our Editorial Board Committee[99] comprises a distinguished panel of international experts in law, technology, and society across different disciplines and domains. I would like to thank Daan Rutten and Charles Dybus from Tilburg University for their help in launching the journal, as well as Roger Brownsword and Mark Coeckelbergh for their invited contributions.

Here we go, let the debate begin!

97 We do seek to also reach policy makers and others. The situation for them might even be worse than for academics.

98 We do need, and solicit, your help though. We need reviewers and editorial board members covering various sub-domains. Please let us know if you want to be of assistance.

99 https://techreg.org/index.php/techreg/about/editorialTeam

02

Technology and Regulation

Law Disrupted, Law Re-Imagined, Law Re-Invented

Roger Brownsword*

Coherence, instrumentalism, technocracy, Rule of Law

roger.brownsword@kcl.ac.uk

This article describes the technological disruption of law and legal reasoning, suggests how law might be re-imagined, and proposes four key elements in its re-invention. Two waves of disruption are identified: one impacting on the content of legal rules and perceptions of their deficiency; a second impacting on our appreciation of technological instruments as tools to be used for regulatory purposes to support or replace legal rules. The suggested re-imagination of law centres on the idea of the regulatory environment. The proposed re-invention of law starts with (i) a fresh understanding of the range of regulatory responsibilities, which shapes (ii) the articulation of the Rule of Law and informs both (iii) a renewal of traditional coherentist thinking and (iv) a reshaping of legal and regulatory institutions.

1. Introduction

This article is about the disruption of law and legal reasoning by new technologies as a result of which, I suggest, there is a need to re-imagine and then to re-invent law. It is about the disruptive impact of new technologies on the traditional content of legal rules, about the way that those associated with the legal and regulatory enterprise reason, about the increasing availability of technological instruments to support, or even supplant, legal rules and, concomitantly, it is about the displacement of human agents from traditional regulatory roles.

The argument is that, in the wake of this disruption, there is a need to re-imagine the field (the regulatory environment) of which legal rules are a part.[1] Instead of thinking exclusively in terms of a certain set of rules and norms (representing 'the law'), it is suggested that we should think of a set of tools that can be employed for regulatory purposes. While some of these tools (such as legal rules) are normative, others (employing, for example, the design of products or processes) are non-normative. While normative instruments always speak to what 'ought' to be done, non-normative instruments at any rate, at the hard end of the spectrum speak only to what 'can' and 'cannot' be done.[2] Finally, it is argued that, if law is to be re-invented, the renewal should be anchored to a new foundational understanding of regulatory responsibilities on which we can draw in order to shape our articulation of the Rule of Law, to revitalise 'coherentist'[3] thinking, and to refashion legal and regulatory institutions.

The article is in four parts. In Part 2, two principal disruptive waves are sketched: while one wave of technological disruption impacts on both the substance of legal rules and the prevailing legal mind-set, the other impacts on our appreciation of rules as just one kind of regulatory instrument. While the first wave has been felt since the early days of industrialisation, it is the second wave that will be critical this century.

There are three elements in Part 3: first, three mind-sets ('coherentist', 'regulatory-instrumentalist', and 'technocratic') generated by these technological disruptions are sketched; secondly, relative to these mind-sets, a short retrospective reflection is offered on Judge Frank Easterbrook's provocative argument that to regroup legal rules relating to modern ICTs as 'the law of cyberspace' would be as unilluminating as the regrouping of legal rules to represent 'the law of the horse';[4] and, thirdly, some initial remarks are made in relation to the question of which mind-set should be engaged and when.

* King's College London and Bournemouth University. This article is largely based on a lecture that was given at the University of Warsaw on November 7, 2018 and, in part, on a lecture that was given in Tilburg (at an event celebrating 25 years of TILT) on January 18, 2019. I am grateful for the comments made and questions asked following both lectures, as well as for feedback from the journal's reviewers. Needless to say, the usual disclaimers apply.

1 See, Roger Brownsword, 'In the Year 2061: From Law to Technological Management' (2015) 7 *Law, Innovation and Technology* 1; 'Field, Frame and Focus: Methodological Issues in the New Legal World' in Rob van Gestel, Hans Micklitz, and Ed Rubin (eds), *Rethinking Legal Scholarship* (Cambridge: Cambridge University Press, 2016) 112; and *Law, Technology and Society Re-imagining the Regulatory Environment* (Abingdon: Routledge, 2019).

2 See, e.g., Roger Brownsword, 'Lost in Translation: Legality, Regulatory Margins, and Technological Management' (2011) 26 *Berkeley Technology Law Journal* 132.

3 By 'coherentist' I mean, roughly speaking, a mind-set that is not only focused on the internal consistency and integrity of a body of doctrine but also that engages with new technologies by asking how that body of doctrine applies to new technological (or other) phenomena. I will elaborate this more fully in Part 3.

4 Frank H. Easterbrook, 'Cyberspace and the Law of the Horse' [1996] *University of Chicago Legal Forum* 207. Although Easterbrook's article is frequently recalled for its provocative claim, most of the paper actually argues, in the

Faced with these disruptions, in Part 4, it is suggested that the required act of re-imagination is to view law and legal rules as one element of a more heterogeneous and more inclusive conception of the regulatory environment specifically, a regulatory environment in which new technologies figure as instruments with regulatory effects. As a first step in this act of re-imagination, it is suggested that we might map the field by reference to (i) the types of measure or instrument employed (rules or non-rule technologies) and (ii) the source of the measure (public or private regulator). Then, with the focus on non-rule technological measures, we can develop the map by reference to (iii) the nature of the technological measure (soft or hard) and (iv) the locus of the intervention (external to agents or internal to agents).

Finally, in Part 5, four main elements of the re-invention of law are proposed. These are (i) a new foundationalist and hierarchical understanding of the range of regulatory responsibilities, where the responsibility to maintain the essential conditions for human social existence (the commons) is prioritised, (ii) a new appreciation of the Rule of Law, (iii) a renewed form of coherentist thinking, and (iv) a refashioning of legal and regulatory institutions.

My conclusion is not that, with law so re-invented, all will go well. In a world of dynamic technological change, maintaining the commons will always be a challenge and discharging our regulatory responsibilities will inevitably be work in progress. Nevertheless, I suggest that the chances of things going well are somewhat better if we do so re-imagine and then re-invent law than if we take no steps in this direction.

2. Law Disrupted

Shortly before Christmas 2018, an unauthorised drone was sighted in the vicinity of the airfield at London Gatwick airport. As a precautionary measure, all flights were suspended and, for two days, the airport was closed.[5] Following this incident, some exhorted the government to change the rules, particularly by providing for an extended drone no-fly zone around airports in response to which, the government announced that the police would be given new powers to tackle illegal drone use,[6] and that the drone no-fly zone would be extended to 3 miles around airports.[7] Others, however, focused, not on the fitness of the rules, but on the possibility of finding a technological solution, ideally one that rendered it impossible in practice for a drone to be flown near an airport (or, failing that, a technology for disabling and bringing down unauthorised drones).[8]

Similarly, in its recent White Paper on the regulation of harmful online content ranging broadly across content that is harmful to national security, to politicians, to children, and so on the UK government has outlined a two-pronged strategy.[9] While one prong of the proposed response focuses on rendering the rules fit for purpose in the digital age (notably by establishing a new statutory duty of care on Internet companies 'to take reasonable steps to keep their users safe and tackle illegal and harmful activity on their services'[10]), the other prong aspires to make 'technology itself [a]...part of the solution'.[11]

In these two responses, focusing on both rule changes and technological solutions, we see the disruption of law represented in two ways. First, there is the thought that the rules are not fit for (regulatory) purpose, this reflecting a sense of the inadequacy of existing legal rules. Secondly, there is the thought that the most effective regulatory response might be to rely on technological instruments rather than rules, this being at odds with the assumption that social order is to be maintained by the use of rules (and, concomitantly, heightening our appreciation of the potential use of both technological instruments other than legal rules and of smart machines rather than human agents). If the former views technology as a disruptive problem, the latter sees technology as part of the solution. If the former is characteristic of disruption that goes back to the early years of industrialisation, the latter is more characteristic of the Millennium.

Law is, thus, disrupted in two waves, one wave impacting on the substance of the rules on which we rely and the other on whether we should rely on rules at all. However, as we will elaborate in the next Part of the article, these disruptions also impact on the way in which we think as lawyers, provoking new framings, new conversations, and new legal and regulatory mind-sets.[12]

2.1 The first disruptive wave

The first wave of disruption causes us to question the adequacy of existing rules of law. In some cases, it is deficiencies in the substance of prevailing legal rules that are highlighted; the rules at issue need to be changed or qualified. In other cases, it is gaps or omissions in the prevailing legal rules that are exposed; new rules need to be introduced. However, in both cases, the essential disruption is that we wonder, as we would now put it, whether the legal rules and principles are fit for purpose.

The disruptive effects of industrialisation on the traditional rules of the criminal law are highlighted by Francis Sayre when, in a seminal article, he remarks on the 'steadily growing stream of offenses punishable without any criminal intent whatsoever.'[13] While this development jars with the traditional idea that there can be no criminal offence without mens rea, the world was changing. As Sayre recognised, the 'invention and extensive use of high-powered automobiles require new forms of traffic regulation;...the growth of modern factories requires new forms of labor regulation; the development of modern building construction and the growth of skyscrapers require new forms of building regulation.'[14] So it was that, in both England and the United States, from the middle of the Nineteenth Century, the courts accepted that, so far as 'public welfare' offences were concerned, it

spirit of Coasean law and economics, for clear rules, for creating property rights where they are needed, and for facilitating the formation of bargaining institutions.

5 See, e.g., BBC News, 'Gatwick airport: How countries counter the drone threat', December 21, 2018, https://www.bbc.co.uk/news/technology-46639099 (last accessed 21 December 2018).

6 See BBC News, Police to get new powers to tackle illegal drone use' January 7, 2019. Available at https://www.bbc.co.uk/news/uk-46787730 (last accessed February 20, 2019).

7 See BBC News (Business), 'Drone no-fly zone to be widened after Gatwick chaos' February 20, 2019. Available at https://www.bbc.co.uk/news/business-47299805 (last accessed February 20, 2019).

8 It has also been reported that the Home Office is testing new counter-drone technologies (see n 7).

9 HM Government, Online Harms White Paper (CP 57, April 2019).

10 'Online Harms White Paper' at p. 42 (para 3.1).

11 'Online Harms White Paper' at p. 6 (para 10) so, for example, at p. 13, para 1.12, we read that it is 'vital to ensure that there is the technology in place to automatically detect and remove terrorist content within an hour of upload, secure the prevention of re-upload and prevent, where possible, new content being made available to users at all.' For the various ways in which the government proposes to encourage the search for technological solutions, see Part 4 of the White Paper.

12 See, Roger Brownsword, 'Law and Technology: Two Modes of Disruption, Three Legal Mind-Sets, and the Big Picture of Regulatory Responsibilities' (2018) 14 *Indian Journal of Law and Technology* 1; and Law, Technology and Society Re-imagining the Regulatory Environment (Abingdon: Routledge, 2019) Chs 8-12.

13 F.B Sayre, 'Public Welfare Offences' (1933) 33 *Columbia Law Review* 55, at 55.

14 Sayre (n 13) at 68-69.

was acceptable to dispense with proof of intent or negligence.[15] If the food sold was adulterated, if vehicles did not have lights that worked, if waterways were polluted, and so on, sellers and employers were simply held to account. For the most part, this was no more than a tax on business; it relieved the prosecutors of having to invest time and resource in proving intent or negligence; and, as Sayre reads the development, it reflected 'the trend of the day away from nineteenth century individualism towards a new sense of the importance of collective interests.'[16]

A somewhat similar story of disruption can be told in relation to the rules of tort law. There, the key developments involve adjustments to the cornerstone idea of fault-based liability.[17] As Geneviève Viney and Anne Guégan-Lécuyer put it, a tort regime 'which seemed entirely normal in an agrarian, small-scale society, revealed itself rather quickly at the end of the nineteenth century to be unsuitable.'[18] Accordingly, stricter forms of liability were needed to assist claimants who had been exposed to unacceptable forms of risk. However, at the same time, it was necessary to introduce immunities in order to shield nascent enterprises and to maintain an environment that does not discourage innovation.[19]

In the case of contract law, the key moments of disruption start with a shift from a 'subjective' consensual model of agreement to an 'objective' approach. The idea that contractors have to be subjectively ad idem, actually to have agreed on the terms and conditions of the transaction, hampered enterprises that needed to limit their liabilities associated with new transportation technologies. In the common law jurisprudence, this shift is epitomised by Mellish LJ's direction to the jury in *Parker v South Eastern Railway Co*,[20] where the legal test is said to be not so much whether a customer actually was aware of the terms and had agreed to them but whether the railway company had given reasonable notice.[21] About a hundred years later, we come to a second moment of disruption when, with the development of a mass consumer market for new technological products (cars, televisions, kitchen appliances, and so on), it was necessary to make a fundamental correction to the traditional value of 'freedom of contract' in order to protect consumers against the small print of suppliers' standard terms and conditions. Finally, although the potentially disruptive effects of online environments for commerce and contracting were resisted, it remains an open question whether the law can continue to treat contracts that are made using new transactional technologies as if they were traditional offline, non-automated, non-self-enforcing transactions.[22]

What we see across these developments is a pattern of disruption to legal doctrines that were organically expressed in smaller-scale non-industrialised communities communities where horses, not machines, did the heavy work. Here, the legal rules presuppose very straightforward ideas about holding to account (moreover, holding *personally* to account) those who engage intentionally in injurious or dishonest acts, about expecting others to act with reasonable care, and about holding others to their word. Once new technologies disrupt these ideas, we see the move to strict or absolute criminal liability without proof of intent, to tortious liability without proof of fault, to vicarious liability, and to contractual liability (or limitation of liability) without proof of actual intent, agreement or consent. Moreover, these developments signal a doctrinal bifurcation,[23] with some parts of criminal law, tort law and contract law resting on traditional principles (and representing, so to speak, 'real' crime, tort and contract) while others deviate from these principles as necessary adjustments or corrections are made.

More recently, we find a number of landmark cases in which the development or application of a new technology has exposed gaps or omissions in the law. For example, in the 1970s, Patrick Steptoe and Robert Edwards pioneered the development of the technique of in vitro fertilisation (IVF), famously leading to the birth of Louise Brown in 1978. Although the collaboration between Steptoe and Edwards did not involve any unlawful activity as such, the use of IVF was not explicitly legally authorised and, following the successful use of IVF, the Warnock Committee was set up to make recommendations concerning both assisted conception and the use of human embryos for research. In due course, the Human Fertilisation and Embryology Act, 1990, was put in place. This new legal framework set out the groundrules for the provision of, and access to, IVF services as well as for licensing research using human embryos. Similarly, various technological developments have provoked the creation of new offences to deal with a range of matters from human reproductive cloning to cybercrime. The development of computers necessitated setting out a legal framework for the processing of personal data; and there has been sui generis gap-filling and stretching of IP law to cover such matters as databases, software, and integrated circuits. What is distinctive about this kind of disruption is not so much that there are additions to the legal rule-book but that these responses are typically bespoke, tailored and in a legislative form; and, critically, the regulatory mind-set that directs these responses is quite different to traditional coherentist patterns of thought. Because this is a matter to which we will return in Part 3 of the article, we can leave it at that for the moment.

2.2 The second disruptive wave

The focus of the second disruptive wave is not on the deficient content of prevailing legal rules, or on gaps, but on the availability of new technological instruments that can be applied for regulatory purposes. The response to such disruption is not that some rule changes or new rules are required but that the use of rules is not necessarily the most effective way of achieving the desired regulatory objective.

15 So far as the development in English law is concerned, illustrative cases include R v Stephens LR 1 QB 702 (1866); Hobbs v Winchester [1910] 2 KB 471; and Provincial Motor Cab Co v Dunning [1909] 2 KB 599.

16 Sayre (n 13) at 67.

17 See Miquel Martin-Casals (ed), *The Development of Liability in Relation to Technological Change* (Cambridge: Cambridge University Press, 2010).

18 Geneviève Viney and Anne Guégan-Lécuyer, 'The Development of Traffic Liability in France' in Martin-Casals (n 17) 50.

19 For example, in the United States, the interests of the farming community were subordinated to the greater good promised by the development of the railroad network: see Morton J. Horwitz, *The Transformation of American Law* 1780-1860 (Cambridge, Mass.: Harvard University Press, 1977).

20 (1877) 2 C.P.D. 416.

21 Nb, too, Stephen Waddams, *Principle and Policy in Contract Law (Cambridge*: Cambridge University Press, 2011) at 39, pointing out that the emphasis of Bramwell LJ's judgment in Parker is 'entirely on the reasonableness of the railway's conduct of its business and on the unreasonableness of the customers' claims; there is no concession whatever to the notion that they could only be bound by their actual consent.' For a fine example of principled contractual thinking coming into tension with regulatory reasoning, see Catharine MacMillan, 'The Mystery of Privity: Grand Trunk Railway Company of Canada v Robinson (1915)' (2015) 65 *University of Toronto Law Journal* 1.

22 See, e.g., Roger Brownsword, 'The E-Commerce Directive, Consumer Transactions, and the Digital Single Market: Questions of Regulatory Fitness, Regulatory Disconnection and Rule Redirection' in Stefan Grundmann (ed), *European Contract Law in the Digital Age* (Cambridge: Intersentia, 2017) 165.

23 As recognised, for example, in the Canadian Supreme Court case of R. v. Sault Ste. Marie [1978] 2 S.C.R. 1299, at 1302-1303.

Already, this presupposes a disruption to traditional patterns of legal thinking that is to say, it presupposes a regulatory-instrumentalist and purposive mind-set and a willingness to think about turning to architecture, design, coding, AI, and the like as a regulatory tool. Arguably, we can find such a willingness as soon as people fit locks on their doors. However, the variety and sophistication of the technological instruments that are available to regulators today is strikingly different to the position in both pre-industrial and early industrial societies. In particular, there is much more to technological management than traditional target-hardening: the management involved might—by designing products and places, or by coding products and people—disable or exclude potential wrongdoers as much as harden targets or immunise potential victims; and, there is now the prospect of widespread automation that takes humans altogether out of the regulatory equation. Crucially, with a risk management approach well-established, regulators now find that they have the option of responding by employing various technological instruments rather than rules. This is the moment when, so to speak, we see a very clear contrast between the legal and regulatory style of the rule-governed East coast (whether traditional or progressive) and the technologically-managed style of the West coast.[24]

In the wake of this second disruptive wave, the take-up of technological tools can be charted on a spectrum running from soft to hard.[25] At the soft end of the spectrum, the technologies are employed in support of the legal rules. For example, the use of surveillance technologies and/or identification technologies signals that rule-breaking is more likely to be detected; other things being equal, compliance with the rules is assisted and encouraged; but the strategy is still rule-based and the practical option of non-compliance remains. By contrast, at the hard end of the spectrum, the focus and the ambition are different. Here, measures of 'technological management' focus on limiting the practical (not the paper) options of regulatees;[26] and, whereas legal rules back their prescriptions with *ex post* penal, compensatory, or restorative measures, the focus of technological management is entirely *ex ante*, aiming to anticipate and prevent wrongdoing rather than punish or compensate after the event. Albeit a measure for road safety rather than crime control, this is how we should interpret the recent EU proposal to require that all new cars should be fitted with devices that ensure that vehicles comply with speed limits.[27]

Elsewhere, we see the search for technological solutions in relation to the protection of both intellectual property rights (qua digital rights management) and privacy.[28] Granted, a good deal of the effort to find such solutions comes from private corporations who deploy technological measures that have the desired regulatory and risk-managing effects. To this extent, these parties act as regulators, albeit not as public regulatory bodies. It is also true that public regulators for example, in relation to the regulation of online content may direct or encourage private parties to develop technological solutions rather than invest in and impose their own technological measures. During the second wave of disruption, all parties who are in a position to 'regulate' begin to appreciate the possibilities given by new technological tools.

To elaborate on these latter examples, with the development of computers and then the Internet and World Wide Web, supporting a myriad of applications, it is clear that, when individuals operate in online environments, they are at risk in relation to both their 'privacy' and the fair processing of their personal data. Initially, regulators assumed that 'transactionalism' would suffice to protect individuals: in other words, it was assumed that, unless the relevant individuals agreed to, or consented to, the processing of their details, it would not be lawful. However, once it was evident that consumers in online environments routinely signalled their agreement or consent in a mechanical way, without doing so on a free and informed basis, a more robust risk-management approach invited consideration. Such an approach might still be rule-based (probably with the reasonableness of online business practice setting the standard), but the management might also be technological. In other words, once we are thinking about the protection of the autonomy of internet-users or about the protection of their privacy, why not also consider the use of technological instruments in service of the regulatory objectives?

Indeed, in Europe, this kind of thinking resonates with what we find in the General Data Protection Regulation (GDPR)[29] and, similarly, in Article 13 (now renumbered 17) of the EU Copyright Directive (where content recognition technologies and further development of such technologies are treated as central to cooperative arrangements between copyright holders and information society service providers).[30] While talk of 'privacy enhancing technologies' and 'privacy by design' has been around for some time,[31] in the GDPR we see that this is more than talk; it is not just that the regulatory discourse is more technocratic, there are signs that the second disruption is beginning to impact on regulatory practice—although how far this particular impact will penetrate remains to be seen.[32]

24 Seminally, see Lawrence Lessig, *Code and Other Laws of Cyberspace* (New York: Basic Books, 1999). See, too, Roger Brownsword, 'Code, Control, and Choice: Why East is East and West is West' (2005) 25 *Legal Studies* 1.

25 See, e.g., Pat O'Malley, 'The Politics of Mass Preventive Justice' in Andrew Ashworth, Lucia Zedner, and Patrick Tomlin (eds), *Prevention and the Limits of the Criminal Law* (Oxford University Press, 2013) 273.

26 See, e.g., Roger Brownsword, 'Law, Liberty and Technology' in Roger Brownsword, Eloise Scotford, and Karen Yeung (eds), *The Oxford Handbook of Law, Regulation and Technology* (Oxford: Oxford University Press, 2017) 41.

27 See, Graeme Paton, 'Automatic speed limits planned for all new cars' *The Times*, March 27, p 1.

28 Compare, Lee A. Bygrave, 'Hardwiring Privacy' in Roger Brownsword, Eloise Scotford, and Karen Yeung (n 26), 754 , 755. Here, Bygrave says that, in the context of the design of information systems, the assumption is that, by embedding norms in the architecture, there is 'the promise of a significantly increased ex ante application of the norms and a corresponding reduction in relying on their application *ex post facto*.'

29 Regulation (EU) 2016/679. See, e.g., Recital 78 which enjoins data controllers to take 'appropriate technical and organisational measures' to ensure that the requirements of the Regulation are met; and similarly, in the body of the GDPR, see Article 25 (concerning data protection by design and by default).

30 European Commission, Proposal for a Directive of the European Parliament and of the Council on copyright in the Digital Single Market, COM(2016) 593 final, 2016/0280(COD) (Brussels, 14.9.2016).

31 See, Bygrave (n 28); Ann Cavoukian, Privacy by Design: The Seven Foundational Principles (Information and Privacy Commissioner of Ontario, 2009, rev ed 2011) (available at https://www.ipc.on.ca/images/Resources/7foundationalprinciples.pdf) (last accessed February 5, 2018). For a recent review of the use, development and limits of a range of PETs, see The Royal Society, Protecting privacy in practice (London, March 2019). One of the recommendations made in this report is that government and regulators should 'support organisations to become intelligent users of PETs'. So, for example, 'the Information Commissioner's Office (ICO) should provide guidance about the use of suitably mature PETs to help organisations minimise risks to data protection, and this should be part of the ICO's Data Protection Impact Assessment guidelines. Such guidance would need to cover how PETs fit within an organisation's overall data governance infrastructure, since the use of PETs in isolation is unlikely to be sufficient' (at 7).

32 Bygrave (n 28) argues, at 756, that, despite explicit legal backing, 'the privacy-hardwiring enterprise will continue to struggle to gain broad traction.' Most importantly, this is because this enterprise 'is at odds with powerful

This evolution in regulatory thinking is not surprising. Having recognised the limited fitness of traditional legal rules, and having taken a more regulatory approach, the next step is to think not just in terms of risk assessment and risk management but also to be mindful of the technological instruments that increasingly become available for use by regulators. In this way, the regulatory mind-set is focused not only on the risks to be managed but also how best to manage those risks (including making use of technological tools).[33]

3. The Legal Mind-Set Disrupted

It will be recalled that one of the impacts of the first wave of disruption is to destabilise the traditional coherentist mind-set the challenge comes from a mind-set the logic of which is altogether more purposive and regulatory-instrumentalist. This disruptive effect is compounded by the second wave of disruption when regulatory-instrumentalism is taken in a more technocratic direction. With each mind-set, there are different questions that are focal, different framings, and different conversations that ensue.

Elaborating these disruptive impacts, there are three elements in this part of the article. First, there is a sketch of the three legal and regulatory mind-sets that are central to the narrative: namely, the coherentist, the regulatory-instrumentalist, and the technocratic. Secondly, relative to these mind-sets, I offer a retrospective comment on Judge Frank Easterbrook's famous assertion that creating a dedicated 'law of cyberspace' would be as mindless and inappropriate as recognising a 'law of the horse'.[34] Although we might quickly dismiss Easterbrook's intervention as seriously misreading the runes or as underestimating the significance of the regulatory activity at the technological nodes of interest, I suggest that his view is best regarded as a textbook expression of traditional coherentist thinking. Thirdly I will present some initial reflections on the question of which mind-set should be engaged and when. This is an important question and one to which we will return in Part 5.

3.1 The three mind-sets

In what follows, we present three thumbnail sketches of the legal and regulatory mind-sets to which we have referred: the coherentist, the regulatory-instrumentalist, and the technocratic.

Coherentism

Coherentism is defined by four characteristics. First, what matters above all is the integrity and internal consistency of legal doctrine. This is viewed as desirable in and of itself. Secondly, coherentists are not concerned with the fitness of the law for its regulatory purpose. Thirdly, coherentists approach new technologies by asking how they fit within existing legal categories (and then try hard to fit them in). Fourthly, coherentists believe that legal reasoning should be anchored to guiding general principles. Coherentism is, thus, the natural language of litigators and judges, who seek to apply the law in a principled way.[35]

In the context of rapidly emerging technologies, it is worth lingering over the coherentist tendency to ask not whether the prevailing (and disrupted) rules are fit for purpose but how new phenomena can be fitted into traditional classification schemes or how they comport with general principles of law.

For coherentists, the focus is on the recognised legal concepts, categories and classifications;[36] and this is accompanied by a certain reluctance to abandon these concepts, categories and classifications with a view to contemplating a bespoke response. For example, rather than recognise new types of intellectual property, coherentists will prefer to tweak existing laws of patents and copyright.[37] Similarly, in transactions, coherentists will want to classify e-mails as either instantaneous or non-instantaneous forms of communication (or transmission),[38] they will want to apply the standard formation template to online shopping sites, they will want to draw on traditional notions of agency in order to engage electronic agents and smart machines,[39] and they will want to classify individual 'prosumers' and 'hobbyists' who buy and sell on new platforms (such as platforms that support trade in 3D printed goods) as either business sellers or consumers.[40] As the infrastructure for transactions becomes ever more technological the tension between this strand of coherentism and regulatory-instrumentalism becomes all the more apparent.[41] In sum, coherentism presupposes a world of, at most, leisurely change. It belongs to the age of the horse, not to the age of the autonomous vehicle.

Regulatory-Instrumentalism

In contrast with coherentism, regulatory-instrumentalism is defined by the following three features. First, it is not concerned with the internal consistency of legal doctrine. Secondly, it is entirely focused on whether the law is instrumentally effective in serving specified regulatory purposes and policies. Thirdly, regulatory instrumentalism has no reservation about enacting new bespoke laws if this is an effective and efficient response to a question raised by new technologies. Regulatory-instrumentalism is, thus, the natural language of legislators and policy-makers.

business and state interests, and simultaneously remains peripheral to the concerns of most consumers and engineers' (ibid).

33 Compare Colin Gavaghan, 'Lex Machina: Techno-regulatory Mechanisms and "Rules by Design"' (2017) 15 *Otago Law Review* 123, 145 concluding that techno-regulatory mechanisms 'are already widespread and, likely to become more so as our lives become more urbanized and technologized.'

34 Easterbrook (n 4).

35 For a somewhat similar view, presented as a 'legalistic approach' to emerging technologies, see Nicolas Petit, 'Law and Regulation of Artificial Intelligence and Robots: Conceptual Framework and Normative Implications': available at https://papers.ssrn.com/sol3/papers.cfm?abstract_id=2931339 (last accessed February 17, 2018).

36 See, e.g., the excellent analysis in Shawn Bayern, Thomas Burri, Thomas D. Grant, Daniel M. Häusermann, Florian Möslein, and Richard Williams, 'Company Law and Autonomous Systems: A Blueprint for Lawyers, Entrepreneurs, and Regulators' (2017) 9 *Hastings Science and Technology Law Journal* 135, where company structures that are provided for in US, German, Swiss, and UK law are reviewed to see whether they might plausibly act as a host for autonomous systems that provide a service (such as file storage, file retrieval and metadata management).

37 Compare the analysis of multi-media devices in Tanya Aplin, *Copyright Law in the Digital Society: the Challenges of Multimedia* (Oxford: Hart, 2005).

38 See, e.g., Andrew Murray, 'Entering into Contracts Electronically: the Real WWW' in Lilian Edwards and Charlotte Waelde (eds), *Law and the Internet: A Framework for Electronic Commerce* (Oxford: Hart, 2000) 17; and Eliza Mik, 'The Effectiveness of Acceptances Communicated by Electronic Means, Or – Does the Postal Acceptance Rule Apply to Email?' (2009) 26 *Journal of Contract Law* 68 (concluding that such classificatory attempts should be abandoned).

39 Compare, e.g., Emily Weitzenboeck, 'Electronic Agents and the Formation of Contracts' (2001) 9 *International Journal of Law and Information Technology* 204.

40 Compare e.g., Christian Twigg-Flesner, 'Conformity of 3D Prints—Can Current Sales Law Cope?' in R. Schulze and D. Staudenmayer (eds), *Digital Revolution: Challenges for Contract Law in Practice* (Baden-Baden: Nomos, 2016) 35.

41 For insightful discussion of the proposed B2B platform Regulation, see Christian Twigg-Flesner, 'The EU's Proposals for Regulating B2B Relationships on online platforms Transparency, Fairness and Beyond' (2018) 7 *Journal of European Consumer and Market Law* 222.

The regulatory mind-set is, at all stages, instrumental and instrumentally rational. The question is: what works, what will serve certain specified purposes? When a regulatory intervention does not work, it is not enough to restore the status quo; rather, further regulatory measures should be taken, learning from previous experience, with a view to realising the regulatory purposes more effectively. Hence, the purpose of the criminal law is not simply to respond to wrongdoing (as corrective justice demands) but to reduce crime by adopting whatever measures of deterrence promise to work.[42] Similarly, in a safety-conscious community, the purpose of tort law is not simply to respond to wrongdoing but to deter practices and acts where agents could easily avoid creating risks of injury and damage. For regulatory-instrumentalists, the path of the law should be progressive: we should be getting better at regulating crime and improving levels of safety.[43]

According to Edward Rubin, regulatory-instrumentalism is displacing a coherentist approach.[44] Thus, in the modern administrative state, the 'standard for judging the value of law is not whether it is coherent but rather whether it is effective, that is, effective in establishing and implementing the policy goals of the modern state.'[45] Certainly, one of the striking features of the European Union has been the single market project, a project that the Commission has pursued in a spirit of conspicuous regulatory-instrumentalism. Here, the regulatory objectives are: (i) to remove obstacles to consumers shopping across historic borders; (ii) to remove obstacles to businesses (especially small businesses) trading across historic borders; and (iii) to achieve a high level of consumer protection. In order to realise this project, it has been essential to channel the increasing number of member states towards convergent legal positions.

As the single market project has evolved into the digital Europe project, the Commission's regulatory-instrumentalist mind-set remains perfectly clear. As the Commission puts it:

> The pace of commercial and technological change due to digitalisation is very fast, not only in the EU, but worldwide. The EU needs to act now to ensure that business standards and consumer rights will be set according to common EU rules respecting a high-level of consumer protection and providing for a modern business friendly environment. It is of utmost necessity to create the framework allowing the benefits of digitalisation to materialise, so that EU businesses can become more competitive and consumers can have trust in high-level EU consumer protection standards. By acting now, the EU will set the policy trend and the standards according to which this important part of digitalisation will happen.[46]

In this context, coherentist thoughts about tidying up and standardising the lexicon of the consumer acquis, or pushing ahead with a proposed Common European Sales Law,[47] or codifying European contract law drop down the list of priorities. For regulatory-instrumentalists, when we question the fitness of the law, we are not asking whether legal doctrine is consistent, we are asking whether it is fit for delivering the regulatory purposes.

Last but not least, I take it to be characteristic of the regulatory-instrumentalist mind-set that the thinking becomes much more risk-focused. In the criminal law and in torts, the risks that need to be assessed and managed relate primarily to physical and psychological injury and to damage to property and reputation; in contract law, it is economic risks that are relevant. So, for example, we see in the development of product liability a scheme of acceptable risk management that responds to the circulation of products (such as cars or new drugs) that are beneficial but also potentially dangerous. However, this response is still in the form of a revised *rule* (it is not yet technocratic); and it is still in the nature of an *ex post* correction (it is not yet *ex ante* preventive). Nevertheless, it is only a short step from here to a greater investment in *ex ante* regulatory checks (for food and drugs, chemicals, and so on) and to the use of new technologies as preventive regulatory instruments. In other words, it is only a short step from risk-managing regulatory-instrumentalist thinking to a more technocratic mind-set.

Technocratic

The third mind-set, evolving from a regulatory-instrumentalist view, is one that is technocratic. In response to the demand that 'there needs to be a law against this', the technocratic mind-set, rather than drafting new rules, looks for technological solutions. Such a mind-set is nicely captured by Joshua Fairfield when, writing in the context of non-negotiable terms and conditions in online consumer contracts, he remarks that 'if courts [or, we might say, the rules of contract law] will not protect consumers, robots will.'[48]

We should not assume, however, that technocratic solutions will be accepted without resistance. For example, in the USA, a proposal to design vehicles so that cars were simply immobilised if seat belts were not worn was eventually rejected.[49] Although the (US) Department of Transportation estimated that the so-called interlock system would save 7,000 lives per annum and prevent 340,000 injuries, 'the rhetoric of prudent paternalism was no match for visions of technology and "big brotherism" gone mad'.[50] Taking stock of the legislative debates of the time, Jerry Mashaw and David Harfst remark:

> Safety was important, but it did not always trump liberty. [In the safety lobby's appeal to vaccines and guards on machines] the freedom fighters saw precisely the dangerous, progressive logic of regulation that they abhorred. The private passenger car was not a disease or a workplace, nor was it a common carrier. For Congress in 1974, it was a private space.[51]

42 Compare David Garland, *The Culture of Control: Crime and Social Order in Contemporary Society* (Oxford: Oxford University Press, 2001); and Amber Marks, Benjamin Bowling, and Colman Keenan, 'Automatic Justice? Technology, Crime, and Social Control' in Brownsword, Scotford, and Yeung (n 26) 705.

43 The parallel development of a risk-management ideology in both criminal law and tort is noted by Malcolm Feeley and Jonathan Simon, 'Actuarial Justice: The Emerging New Criminal Law' in David Nelken (ed), *The Futures of Criminology* (London: Sage, 1994) 173.

44 Edward L. Rubin, 'From Coherence to Effectiveness' in Rob van Gestel, Hans-W Micklitz, and Edward L. Rubin (eds), *Rethinking Legal Scholarship* (New York: Cambridge University Press, 2017) 310, 311.

45 Rubin (n 44) 328.

46 European Commission, Communication from the Commission to the European Parliament, the Council and the European Economic and Social Committee, 'Digital contracts for Europe—Unleashing the potential of e-commerce' COM(2015) 633 final (Brussels, 9.12.2015), 7.

47 Despite a considerable investment of legislative time, the proposal was quietly dropped at the end of 2014. This also, seemingly, signalled the end of the project on the Common Frame of Reference in which, for about a decade, there had been a huge investment of time and resource.

48 Joshua Fairfield, 'Smart Contracts, Bitcoin Bots, and Consumer Protection' (2014) 71 *Washington and Lee Law Review Online* 36, 39.

49 See, Jerry L. Mashaw and David L. Harfst, *The Struggle for Auto Safety* (Cambridge, Mass.: Harvard University Press, 1990) Chapter 7.

50 Mashaw and Harfst (n 49) 135.

51 Mashaw and Harfst (n 49) 140.

Today, similar debates might be had about the use of mobile phones by motorists. There are clear and dramatic safety implications but many drivers persist in using their phones while they are in their cars. If we are to be technocratic in our approach, perhaps we might seek a design solution that disables phones within cars, or while the user is driving. However, once automated vehicles relieve 'drivers' of their safety responsibilities, it seems that the problem will drop away—rules that penalise humans who use their mobile phones while driving will become redundant; humans will simply be transported in vehicles and the one-time problem of driving while phoning will no longer be an issue.

While the contrast between a technocratic approach and coherentism is sharp the former not being concerned with doctrinal integrity and not being entirely focused on restoring the status quo prior to wrongdoing the contrast with regulatory-instrumentalism is more subtle. For both regulatory-instrumentalists and technocrats the law is to be viewed in a purposive and policy-orientated way; and, indeed, as we have said, the latter can be regarded as a natural evolution from the former. In both mind-sets, it is a matter of selecting the tools that will best serve desired purposes and policies; and, so long as technologies are being employed as tools that are designed to assist with a rule-based regulatory enterprise as is the case with the examples of drones at Gatwick airport and harmful online content that we mentioned earlier in the article the technocratic approach might be viewed as merely an offshoot from the stem of regulatory-instrumentalism. However, once technocrats contemplate interventions at the hard end of the spectrum, their thinking departs from order based on rules to one based on technological management, from correcting and penalising wrongdoing to preventing and precluding wrongdoing, and from reliance on rules and standards to employing technological solutions. At this point, the technocratic mind-set reflects a new paradigm.

3.2 Disruption denied and the horse that bolted: was Easterbrook wrong?

Famously, Judge Frank Easterbrook, speaking at an early conference on the 'Law of Cyberspace', argued that 'the best way to learn the law applicable to specialized endeavors is to study general rules'.[52] Hence, Easterbrook claimed, to present a course on the 'Law of Cyberspace' would be as misconceived and unilluminating as to present a course on 'The Law of the Horse'. It would be 'shallow' and it would 'miss unifying principles'.[53] Rather, the better approach is 'to take courses in property, torts, commercial transactions, and the like....[For only] by putting the law of the horse in the context of broader rules about commercial endeavors could one really understand the *law* about horses.'[54] Nevertheless, the law of cyberspace was a horse that was destined to bolt. Easterbrook's doubts notwithstanding, courses and texts on 'cyberlaw', or 'Internet law', or 'e commerce', or the like, abound and few would deny that they have intellectual integrity and make pedagogic sense. Similarly, research centres that are dedicated to the study of cyberlaw (or law and technology more generally) have mushroomed and are seen as being in the vanguard of legal scholarship.

That said, was Easterbrook wrong?[55] As we have said, history has proved that Easterbrook was wrong insofar as he was predicting that cyberlaw (or other law/technology projects) would have no future. However, there are also reasons for thinking that Easterbrook was wrong in supposing that the general principles of property, contract, and tort, and the like would represent the key legal material at the new technological nodes of interest. While some of the early case-law on disputes concerning the Internet and on questions provoked by developments in human genetics might have encouraged this view, it is now clear especially so in Europe that bespoke legislation is being put in place to regulate the relevant technologies and their applications.

Nevertheless, to the extent that Easterbrook was expressing a preference for a pedagogic strategy that brings general rules to bear on a range of facts and phenomena (including cyber phenomena), rather than a strategy that isolates cyber phenomena and then assembles the relevant law (both general and particular), his view should not be lightly dismissed. So viewed, the merits of his position hinge on the criteria that we take to be critical for determining the credentials of these rival pedagogic strategies. For example, if we take the criteria to be pedagogic economy, efficiency, and effectiveness, we might think that it is not so clear that Easterbrook was categorically wrong. Indeed, we might think that one of the strengths of Easterbrook's position is that it stands firm in insisting that students should be taught to think in the way that lawyers traditionally think: namely, figuring out how new fact-situations and phenomena fit with general legal rules and principles. Moreover, even if the cyberlaw horse has bolted, many lawyers persist in engaging with new technologies by approaching them in just the way that Easterbrook recommends for example, a common conversation, after blockchain, is whether smart contract applications will be recognised by judges as equivalent to fiat contracts.[56] In other words, coherentist conversations persist. Nevertheless, this supposed strength of Easterbrook's view holds good only so long as what is involved in 'thinking like a lawyer' and what it is to 'really understand the *law*' are unproblematic. Once these desiderata are problematised, Easterbrook's position is open to the objection that it directs the attention of students away from what now really matters, namely the systematically disruptive effects of technology on the law. To fail to foreground such disruption is to fail to understand the relevance and role of the law in a community where processes are increasingly automated and where relations between humans are increasingly mediated and managed by emerging technologies.

Elsewhere, I have suggested that, in our technological time, there are three key questions to be included in the curriculum.[57] These questions are expressed in relation to the teaching of, and curriculum for, the law of contract. However, suitably redrafted, they could be expressed for any of the traditional courses of law that Easterbrook

52 Easterbrook (n 4) 207.
53 Easterbrook (n 4) 207.
54 Easterbrook (n 4) 208.
55 For some responses to Easterbrook, see, e.g., See Lawrence Lessig, 'The Law of the Horse: What Cyberlaw Might Teach' (1999) 113 *Harvard Law Review* 501, and *Code and Other Laws of Cyberspace* (New York: Basic Books, 1999); Andrew Murray, *The Regulation of Cyberspace* (Abingdon: Routledge-Cavendish, 2007) Ch 1, and 'Looking back at the law of the horse: why cyberlaw and the rule of law are important' (2013) 10 *SCRIPTed* 310; and, implicitly, Chris Reed, 'Why judges need jurisprudence in cyberspace' (2018) 38 *Legal Studies* 263.
56 See, e.g., Roger Brownsword, 'Regulatory Fitness: Fintech, Funny Money, and Smart Contracts' (2019) *European Business Organization Law Review* 1-23 DOI 10.1007/s40804-019-00134-2, and. 'Smart Contracts: Coding the Transaction, Decoding the Legal Debates' in Philipp Hacker, Ioannis Lianos, Georgios Dimitropoulos, and Stefan Eich (eds) *Regulating Blockchain: Techno-Social and Legal Challenges* (Oxford: Oxford University Press, 2019) 311.
57 See Roger Brownsword, 'Teaching the Law of Contract in a World of New Transactional Technologies' in Warren Swain and David Campbell (eds), *Reimagining Contract Law Pedagogy: A New Agenda for Teaching* (Legal Pedagogy) (Abingdon: Routledge, 2019) 112.

has in mind. The three questions are as follows: First, how does the law of contract fit in to the wider regulatory environment for transactions? Secondly, as new transactional technologies become available and are taken up, should we try, like 'coherentists', to fit these developments into the existing body of doctrine or should we think about such matters in a more 'regulatory-instrumental' way? Thirdly, what should we make of the possibility of regulatory restrictions or requirements being, so to speak, 'designed into' the emerging technological platforms or infrastructures for contracts? In other words, what should we make of the 'technological management' of transactions?[58]

To ask these questions, we have to understand that law has been disrupted by new technologies. We have to understand that the context in which law operates is one in which legal rules co-exist with technological instruments that support those rules but that also might supplant and supersede such rules. We also have to understand that the traditional coherentist mind-set that is characteristic of court-centred legal thinking has been disrupted by technological developments that reach back into the Nineteenth Century and that it continues to be disrupted by the development, inter alia, of modern information and communication technologies.

Accordingly, in retrospect, what is wrong with Easterbrook's approach is not so much that he defaults to a coherentist mind-set but that he seems to be either unaware of the disruptive effects of technology on the law or thinks that such disruption is unimportant. However, to put law and legal thinking in its modern context, to 'really understand the *law*', it is essential to step outside such a mind-set. Only then is it possible to recognise the extent of the disruption wrought by new technologies and, concomitantly, the significance of legal order. Only then do we begin to understand the uneasy co-existence that might be found in the relationship between law and various tech communities[59] but also *within* different factions of the legal and regulatory community. In sum, the problem with Easterbrook's approach is that it is a denial of (or, in denial about) disruption. While this might be appropriate in the age of the horse, it is not at all appropriate in an age of disruptive cybertechnologies.

3.3 Which mind-set to engage

Given that regulators might frame their thinking in very different ways, does it matter which mind-set they adopt; and, if so, which mind-set should they adopt? When and why should we think like coherentists, when like regulatory-instrumentalists, and when like technocrats?

To illustrate the significance of the regulatory framing, consider the following hypothetical posed by John Frank Weaver:

> [S]uppose the Aeon babysitting robot at Fukuoka Lucle mall in Japan is responsibly watching a child, but the child still manages to run out of the child-care area and trip an elderly woman. Should the parent[s] be liable for that kid's intentional tort?[60]

If we respond to this question (of the parents' liability) with the mind-set of a coherentist, we are likely to be guided by traditional notions of fault, responsibility, causation, and corrective justice. On this view, liability would be assessed by reference to what communities judge to be fair, just and reasonable—and different communities might have different ideas about whether it would be fair, just and reasonable to hold the parents liable in the hypothetical circumstances. By contrast, if we respond like a regulatory-instrumentalist, the thinking is likely to be that before retailers, such as the shop at the mall, are to be licensed to introduce robot babysitters, and before parents are permitted to make use of robocarers, there needs to be a collectively agreed scheme of compensation should something 'go wrong'. On this view, the responsibilities and liabilities of the parents would be determined by the agreed terms of the risk management package. However, we might also imagine a third response, a response of a technocratic nature, seeking to design out the possibility of such an accident. Quite what measures of technological management might be suggested is anyone's guess—perhaps an invisible 'fence' at the edge of the care zone so that children (like supermarket trolleys or golf carts) simply could not stray beyond the limits. However, thinking about the puzzle in this way, the question would be entirely about designing the machines and the space in such a way that (harmful) collisions between children and mall-goers simply could not happen.

Which of these responses is appropriate? On the face of it, coherentism belongs to relatively static and stable communities, not to the dynamic and turbulent technological times of the Twenty-First Century not as a response to unauthorised drones at airports, or to dangerous or distressing online content, or to accidents involving robot carers. Pace Easterbrook, to assume that traditional legal frameworks enable regulators to ask the right questions and answer them in a rational way seems over-optimistic. If we reject coherentism, we will see regulatory-instrumentalism as a plausible default with the option of a technocratic resolution always to be considered.[61] However, there is a concern that regulatory-instrumentalism might tend to 'flatten' decision-making, reducing all conflicts to a balance of interests and replacing respect for fundamental values such as respect for human rights and human dignity with an all-purpose utilitarianism.[62] Moreover, concerns of this kind are amplified by the prospect of the use of technological management. If law is to be re-invented, regulatory-instrumentalism and technological management cannot be the complete answer. Before re-invention, though, we must speak to re-imagination.

4. Law Re-imagined

If technological tools and technologically managed environments are to be a significant part of our regulatory future, then there is a need to re-imagine law: first, setting law in a context that takes full account of the variety of norms that impact on, and influence, human behaviour; and, secondly, placing law in a context that recognises the channelling and constraining effect of technological management. In order to do this, it is suggested that we should broaden the field for juristic inquiry by operating with a notion of the regulatory environment that

58 This possibility (of regulatory effects being coded into software and hardware) is central to Lawrence Lessig's response to Easterbrook, see Lessig (n 55).

59 Compare Karen Yeung, 'Regulation by Blockchain: The Emerging Battle for Supremacy between the Code of Law and Code as Law' (2019) 82 *Modern Law Review* 207.

60 John Frank Weaver, *Robots Are People Too* (Santa Barbara, Ca: Praeger, 2014) 89.

61 For a discussion in point, see David S. Wall, *Cybercrime* (Cambridge: Polity Press, 2007) where a number of strategies for dealing with 'spamming' are considered. As Wall says, if the choice is between ineffective legal rules and a technological fix (filters and the like), then most would go for the latter (at 201).

62 Compare Christophe Geiger, '"Fair Use" through Fundamental Rights: When Freedom of Artistic Expression allows Creative Appropriations and Opens up Statutory Copyright Limitations', Centre for International Intellectual Property Studies, University of Strasbourg, Research Paper No. 2018-09: see, https://www.researchgate.net/publication/328061041_'Fair_Use'_through_Fundamental_Rights_When_Freedom_of_Artistic_Expression_allows_Creative_Appropriations_and_Opens_up_Statutory_Copyright_Limitations.

accommodates both normative rule-based and non-normative technologically-managed approaches. Admittedly, this does not involve the reconceiving of 'law' as such we might continue to conceive of law as a rule-based or norm-based enterprise; and we might continue to conceive of modern legal systems in terms of a conjunction of primary and secondary rules, or as multi-level normative orders, or whatever. In other words, we do not have to concede that 'code' is law, simply that 'code' together with law is part of the regulatory environment. So conceding, the critical correction is to re-imagine law within a regulatory environment that is no longer limited to guidance given by rules or norms.

What would such a regulatory environment look like? Famously, Clifford Shearing and Phillip Stenning highlighted the way in which, at Disney World, the vehicles that carry visitors between locations act as barriers (restricting access).[63] However, theme parks are no longer a special case. We find similar regulatory environments in many everyday settings, where along with familiar laws, rules, and regulations, there are the signs of technological management—for example, we find mixed environments of this kind in homes and offices where air-conditioning and lighting operate automatically, in hotels where the accommodation levels can only be reached by using an elevator (and where the elevators cannot be used and the rooms cannot be accessed without the use of security key cards), and perhaps par excellence in the 'code/space' that we find at airports.[64] On arrival at a modern terminal building, while there are many airport rules to be observed—for example, regulations concerning parking vehicles, smoking in the building, or leaving bags unattended, and so on—there is also a distinctive architecture that creates a physical track leading from check-in to boarding. Along this track, there is nowadays an 'immigration and security zone', dense with identifying and surveillance technologies, through which passengers have little choice other than to pass. In this conjunction of architecture together with surveillance and identification[65] technologies, we have the non-normative dimensions of the airport's regulatory environment—the fact of the matter is that, if we wish to board our plane, we have no practical option other than to follow the technologically managed track. Similarly, if we want to shop at an Amazon Go store, we have no choice other than to subject ourselves to the technologically managed environment of such stores; and, of course, if we visit Amazon or any other platform online, we will probably do so subject to both the specified terms and conditions for access and whatever technological features are embedded in the site.[66]

If we treat the regulatory environment as essentially a signalling and steering environment, then each such environment operates with a distinctive set of regulatory signals that are designed to channel the conduct of regulatees within, so to speak, a regulated sphere of possibility. Of course, one of the benefits of technologies is that they can expand our possibilities; without aircraft, we could not fly. Characteristically, though, the kind of technological management that we are contemplating is one that restricts or reduces existing human possibilities (albeit, in some cases, by way of a trade-off for new possibilities). In other words, while normative regulation is directed at actions that are possible—and that remain possible—technological management engages with spheres of possibility but in ways that restructure those regulatory spaces and redefine what is and is not possible. In technologically managed environments, it is not so much a matter of what we ought or ought not to do but of what we can and cannot do.

This brief introduction to a re-imagined regulatory environment of which law is just one part needs more detail.[67] First, we need to make a few schematic remarks about technological management as a regulatory option before, secondly, offering some initial remarks about the mapping of the field that is to be re-imagined. Here, we propose a general map of the field in which we take our bearings from (i) the types of measure or instrument employed (rules or non-rule technologies) and (ii) the source of the measure (public or private regulator); and, then, we propose a more detailed mapping of the technological part of the field in which we take our bearings from (iii) the nature of the technological measure (soft or hard) and (iv) the locus of the intervention (external to or internal to regulatees). If we were to visualise this map, it would comprise a pair of two-by-two square grids. The first (general) grid would map: (i) rules issued by a public regulator; (ii) rules issued by a private (regulatory) body; (iii) technological measures employed by a public regulator; and (iv) technological measures employed by a private (regulatory) body. The second (technology-specific) grid would map: (i) soft technological measures that are external to regulatees; (ii) soft technological measures that are internal to regulatees; (iii) hard technological measures that are external to regulatees; and, (iv) hard technological measures that are internal to regulatees. In conjunction with the mapping in the first grid, the mapping in the second grid would supply further and better particulars about the types of technological measures employed by public and by private regulators.

4.1 Technological management as a regulatory option

Technological management might employ a variety of measures, including the design of products (such as golf carts or computer hardware and software) and processes (such as the automated production and driving of vehicles, or the provision of consumer goods and services), the design of places (such as the Metro, or theme parks and airports) and spaces (particularly online spaces), and (in future) the design of people. Typically, such measures are employed with a view to managing certain kinds of risks by excluding (i) the possibility of certain actions which, in the absence of this strategy, might be subject only to rule regulation or (ii) human agents who otherwise might be implicated (whether as rule-breakers or as the innocent victims of rule-breaking) in the regulated activities. More-

63 Clifford D. Shearing and Phillip C. Stenning, 'From the Panopticon to Disney World: the Development of Discipline' in Anthony N. Doob and Edward L. Greenspan (eds), *Perspectives in Criminal Law: Essays in Honour of John LL.J. Edwards* (Toronto: Canada Law Book, 1985) 335.

64 Compare James Bridle, *New Dark Age Technology and the End of the Future* (London: Verso, 2018) 37: An airport is a canonical example of what geographers call 'code/space'. Code/spaces describe the interweaving of computation with the built environment and daily experience to a very specific extent: rather than merely overlaying and augmenting them, computation becomes a crucial component of them, such that the environment and the experience of it actually ceases to function in the absence of code.

65 At London Heathrow airport, there is a £50 million project to install facial recognition technology said to be 'the biggest single deployment of biometric technology in the world' that will dispense with the need to show passports and boarding passes along the track. This project is presented as being for the convenience of passengers but it is also designed to increase security. See, Mark Bridge, Graeme Paton, and Daphne Bugler, 'No need for passports as Heathrow goes hi-tech' *The Times*, April 27, 2019, p 1.

66 For an insightful and detailed analysis of the technological management of Facebook's site, see Tomer Shadmy, 'The New Social Contract: Facebook's Community and Our Rights' (2019) 37 *Boston University International Law Journal* (forthcoming).

67 For extended discussion, see Brownsword, 'In the Year 2061: From Law to Technological Management' (2015) 7 *Law, Innovation and Technology* 1; and *Law, Technology and Society: Re-imagining the Regulatory Environment* (Abingdon: Routledge, 2019) Ch 2.

over, technological management might be employed by both public regulators and by private self-regulating agents (such as corporations protecting their IP rights or supermarkets protecting their merchandise and their trolleys).

Schematically, where the use of technological management is available as a regulatory option, the process can be presented in the following terms:

- Let us suppose that a regulator, R, has a view about whether regulatees should be required to, permitted to, or prohibited from doing x (the underlying normative view)
- R's view could be expressed in the form of a rule that requires, permits, or prohibits the doing of x (the underlying rule)
- but, R uses (or directs, or encourages, others to use) technological management rather than a rule
- and R's intention in doing so is to translate the underlying normative view into a practical design that ensures that regulatees do or do not do x (according to the underlying rule)
- the ensuing outcome being that regulatees find themselves in environments where the immediate signals relate to what can and cannot be done, to possibilities and impossibilities, rather than to the underlying normative pattern of what ought or ought not to be done.

This description, however, is highly schematic and what such a process actually amounts to in practice in particular, how transparent the process is, how much debate there is about the underlying normative view and then about the use of technological measures[68] will vary from one context to another, from public to private regulators, between one public regulator and another, and between one private regulator and another.

It also should be emphasised that the ambition of technological management is to replace the rules by controlling the practical options that are open to regulatees. In other words, technological management goes beyond technological assistance in support of the rules. Of course, regulators might first turn to technological instruments that operate in support of the rules. For example, in an attempt to discourage shoplifting, regulators might require or encourage retailers to install surveillance and identification technologies, or technologies that sound an alarm should a person carry goods that have not been paid for through the exit gates. However, this is not yet full-scale technological management. Once such hard technological management is in operation shoppers will find that it is simply not possible to take away goods without having paid for them.

4.2 Mapping the (re-imagined) field

Even if technological disruption is all around them, why should jurists re-imagine law? If their interests are purely doctrinal, if their mindset is purely coherentist, jurists can continue to engage with their traditional puzzles and lines of inquiry. However, to the extent that technological management displaces rules as the regulatory instrument of choice, traditional legal scholarship loses its relevance; rather like those who are experts in a language that is no longer spoken, coherentist lawyers (following Easterbrook) will be experts in a form of social ordering that is no longer practised. Moreover, if jurists hope to be able to contribute to debates about the legitimacy of particular forms of social ordering or particular exercises of power, they need to think beyond coherentism and they need to re-imagine law as one element in a larger configuration of power.

[68] Compare Shadmy (n 66).

A general map

We can concede that jurists will have different cognitive interests and priorities. Nevertheless, assuming a common concern with the who and the how of the exercise of regulatory power, we can propose two sets of features that would give shape to a very general map of the re-imagined field. First, the map should indicate which type of measures or instruments are being used; and, secondly, it should indicate whether the source of the measure or instrument is public (and, typically, top down) or private (and, often, bottom-up).

Employing the first indicator, the map should tell us whether a particular regulatory environment, or a particular regulatory space, is constituted by rules or by non-rule technologies (or, indeed, by some combination of rules and non-rule technologies). Where we are in zones that are regulated by rules, we are in familiar territory; we have centuries of jurisprudential reflection to help us. However, where non-rule technologies are in play, it is a very different story. As Sheila Jasanoff has remarked, even though 'technological systems rival legal constitutions in their power to order and govern society...there is no systematic body of thought, comparable to centuries of legal and political theory, to articulate the principles by which technologies are empowered to rule us.'[69] Accordingly, once we have our most general map in place, we can begin work on a map that will aid our re-imagination of law specifically where non-rule technologies are in play.

Our general map should also tell us whether the source of the measure is public (and, typically, top down) or private (and, often, bottom-up) in other words, whether the regulator is public or private. In much traditional legal scholarship, the focus is on rules that have been promulgated by public law-making bodies. As critics of this approach have objected, this focus neglects the rule-making activities of private bodies. However, even with an expanded focus, we are still presupposing that we are operating in rule-governed zones. Once we move into regulatory spaces where non-rule technologies apply then we are in largely uncharted territory. Even so, it would be surprising if we did not think it important to know whether these technologies have been initiated and are being controlled by public or by private regulators or pursuant to some form of public/private partnership.[70]

That said, it must be admitted that the distinction between public and private is notoriously contestable and that the distinction between top-down and bottom-up regulation is both crude and far from exhaustive. For example, top-down government regulators might enlist the aid of non-governmental intermediaries (such as Internet service providers or platform providers) or they might adopt a co-regulatory approach setting general targets or objectives for regulatees but leaving them to determine how best to comply.[71] With new technologies occupying and disrupting regulatory spaces, regulators need to re-imagine how best to regulate. As Albert Lin says, in his analysis of new distributed innovative technologies (such as DIYbio, 3D printing, and the platforms of the share economy) these new forms of dynamic activity 'confound conventional regulation.'[72] In response, Lin argues, it turns out that '[g]overnance of distributed innovation...

[69] Sheila Jasanoff, *The Ethics of Invention* (New York: W.W. Norton, 2016) 9-10.

[70] For a perceptive commentary on the regulation of smart cities (as an example of technology-reliant public/private partnerships), see Lilian Edwards, 'Privacy, Security and Data Protection in Smart Cities: A Critical EU Law Perspective' (2016) 2 *European Data Protection Law Review* 28.

[71] Compare Julia Black, 'De-centring Regulation: Understanding the Role of Regulation and Self-Regulation in a "Post-Regulatory" World' (2001) 54 *Current Legal Problems* 103.

[72] Albert C. Lin, 'Herding Cats: Governing Distributed Innovation' (2018) 96 *North Carolina Law Review* 945, 965.

must be both distributed and innovative.'[73] There is no one size fits all; and the regulatory environment that is most acceptable and effective is likely to have elements of both top-down and bottom-up approaches together with elements that fit neither of these types.

Nevertheless, as a first cut at re-imagining regulatory spaces, we can work along two axes. On one axis, it is the balance between reliance on rules and reliance on technologies that is measured; and, on the other axis, it is the extent to which regulatory interventions are public and/or top-down or private and/or bottom-up that is measured.

A specific mapping of technological measures

Once we are in areas that are regulated by non-rule technological measures, how should we get our bearings? I suggest again, somewhat tentatively that our map should indicate, first, what the nature of the particular measure is (specifically where it lies on a spectrum between soft and hard intervention) and, secondly, the locus of the intervention (specifically where it lies on a spectrum between external (to regulatees) and internal (to regulatees)).

With regard to the first indicator, we can differentiate between, on the one hand, those technological measures that are merely supportive of existing rules or assistive or advisory in relation to decision-making, and, on the other, measures of technological management proper that aim to eliminate or redefine some part of an agent's practical options. For example, the use of surveillance and identification technologies in the criminal justice system may simply support the rules of the criminal law; and the use of AI in police practice and in criminal justice decision-making may be simply assistive and advisory. By contrast, if vehicles cannot be driven unless seat belts are engaged, we have full-scale technological management.

Some years ago, Mireille Hildebrandt drew a distinction between 'regulative' and 'constitutive' technological features.[74] Whereas the former are in the nature of assistive or advisory technological applications, the latter represent full-scale technological management. By way of an illustrative example, Hildebrandt invites readers to imagine a home that is enabled with a smart energy meter:

> One could imagine a smart home that automatically reduces the consumption of energy after a certain threshold has been reached, switching off lights in empty rooms and/or blocking the use of the washing machine for the rest of the day. This intervention [which is a case of a 'constitutive' technological intervention] may have been designed by the national or municipal legislator or by government agencies involved in environmental protection and implemented by the company that supplies the electricity. Alternatively [this being a case of a 'regulative' technological intervention], the user may be empowered to program her smart house in such a way. Another possibility [again, a case of a 'regulative' technological intervention] would be to have a smart home that is infested with real-time displays that inform the occupants about the amount of energy they are consuming while cooking, reading, having a shower, heating the house, keeping the fridge in function or mowing the lawn. This will allow the inhabitants to become aware of their energy consumption in a very practical way, giving them a chance to change their habits while having real-time access to the increasing eco-efficiency of their behaviour.[75]

Similarly, Pat O'Malley charts the different degrees of technological control on a spectrum running from 'soft' to 'hard' by reference to the regulation of the speed of motor vehicles:

> In the 'soft' versions of such technologies, a warning device advises drivers they are exceeding the speed limit or are approaching changed traffic regulatory conditions, but there are progressively more aggressive versions. If the driver ignores warnings, data—which include calculations of the excess speed at any moment, and the distance over which such speeding occurred (which may be considered an additional risk factor and *thus* an aggravation of the offence)—can be transmitted directly to a central registry. Finally, in a move that makes the leap from perfect detection to perfect prevention, the vehicle can be disabled or speed limits can be imposed by remote modulation of the braking system or accelerator.[76]

Accordingly, whether we are considering smart cars, smart homes, or smart regulatory styles, we need to be sensitive to the way in which the regulatory environment engages with regulatees, whether it directs signals at regulatees enjoining them to act in particular ways, or whether the technology of regulation simply imposes a pattern of conduct upon regulatees irrespective of whether they would otherwise choose to act in the way that the technology now dictates.

At all points on this spectrum, whether the technological instrument is simply advisory and assistive, or becomes a 'nudge' (again running from soft to hard), or becomes a full-blown measure of technological management, we need to be sensitised to the significance of the particular nature of the technological measure.

This takes us to the second specific indicator, the locus of the intervention. For the most part, our assumption is that technological instruments are being embedded in places and spaces in which regulatees find themselves or with which they interact. Hence, we can talk about technologically managed zones or zones that are rule-governed. However, the proliferation of smart portable or wearable devices, together with many other smart products (such as autonomous vehicles) suggests that the relevant regulatory technological features are not so much zones into which human agents enter but extensions of the human agent. Nevertheless, we might persist with the idea that such technological instruments are still external to the agent (qua regulatee). However, with the development of various kinds of augmented reality and implants, the line between external and internal locations becomes more difficult to maintain. As Franklin Foer has suggested, the development of wearables such as 'Google Glass and the Apple Watch [might] prefigure the day when these companies implant their artificial intelligence within our bodies'.[77] In due course, if, in addition to coded spaces and coded products, we have coded human agents (analogous to coded robots), the line between external and internal signalling would have been crossed.

Taking stock, our general map will enable us to identify the type of regulatory measure (rule or non-rule technological) employed together with the source of that measure (public or private); and, where the measure is non-rule technological, our specific map will enable us to identify whether it is a soft or hard intervention and whether the locus is external or internal to regulatees. Even if we are not quite sure how to respond to a particular measure, this initial

73 Lin (n 72) 1011.

74 Mireille Hildebrandt, 'Legal and Technological Normativity: More (and Less) than Twin Sisters' (2008) 12.3 *TECHNE* 169.

75 Hildebrandt (n 74) 174.

76 Pat O'Malley, 'The Politics of Mass Preventive Justice' in Andrew Ashworth, Lucia Zedner, and Patrick Tomlin (eds), *Prevention and the Limits of the Criminal Law* (Oxford University Press 2013) 273, 280.

77 Franklin Foer, *World Without Mind* (London: Jonathan Cape, 2017) 2.

mapping at least helps us to reconstruct our sense of the landscape of law and to grasp how regulatory power is articulated and by whom.

5. Law Re-invented

In this final part of the article, I outline four respects in which law needs to be re-invented. These concern the range of regulatory responsibilities, the Rule of Law, the renewal of coherentist thinking, and the re-designing of legal and regulatory institutions.

5.1 Regulatory responsibilities

We can start by noting two salient features (and striking problems) in relation to current thinking about regulatory responsibilities. The first is the assumption that whatever particular principles or purposes are taken to be guiding, they are in the final analysis reasonably and rationally contestable; and, the second is the ubiquitous engagement in all manner of balancing exercises (between rights, interests, public policy and so on) without any clear sense of there being a hierarchy that guides deciding between conflicting considerations. In short, there is a lack of foundations; and, there is a lack of hierarchy. Accordingly, a priority for the re-invention of law is to restore some order to our understanding of regulatory responsibilities.

In that spirit, I suggest that we frame our thinking by articulating three tiers of regulatory responsibility, the first tier being foundational, and the responsibilities being ranked in three tiers of importance. At the first and most important tier, regulators have a 'stewardship' responsibility for maintaining the pre-conditions for human social existence, for any kind of human social community. I will call these conditions 'the commons'.[78] At the second tier, regulators have a responsibility to respect the fundamental values of a particular human social community, that is to say, the values that give that community its particular identity. At the third tier, regulators have a responsibility to seek out an acceptable balance of legitimate interests. The responsibilities at the first tier are cosmopolitan and non-negotiable (the red lines here are hard); the responsibilities at the second and third tiers are contingent, depending on the fundamental values and the interests recognised in each particular community. Any conflicts between these responsibilities are to be resolved by reference to the tiers of importance: responsibilities in a higher tier always outrank those in a lower tier.

In what follows, I speak briefly to each of these three tiers before returning to the question of which regulatory mind-set should be engaged.

The regulatory responsibility for the commons

It is an article of faith in the medical profession that doctors should, first, do no harm (to their patients). For regulators, the equivalent injunction should be, first, to ensure that no harm is done to the generic conditions that underpin the lives and prospects of their regulatees.

This injunction rests on two simple but fundamental ideas. First, there is the undeniable fact that members of the human species have certain biologically-dictated needs. Most planets will not support *human* life. The conditions on planet Earth are special for *humans*. Secondly, in the current state of the evolution of the species, humans have the capacity for agency understood in a thin sense akin to that presupposed by the criminal law.[79] That is to say, humans have the capacity to pursue various projects and plans whether as individuals, in partnerships, in groups, or in whole communities. Sometimes, the various projects and plans that they pursue will be harmonious; but, often, human agents will find themselves in conflict or competition with one another as their preferences, projects and plans clash. However, before we get to particular projects or plans, before we get to conflict or competition, there needs to be a context in which the exercise of agency is possible. This context is not one that privileges a particular articulation of agency; it is prior to, and entirely neutral between, the particular plans and projects that agents individually favour; the conditions that make up this context are generic to agency itself. In other words, there is a deep and fundamental critical infrastructure, a commons, for any community of agents. It follows that any agent, reflecting on the antecedent and essential nature of the commons must regard the critical infrastructural conditions as special. Indeed, from any practical viewpoint, prudential or moral, that of regulator or regulatee, the protection of the commons must be the highest priority.

Accordingly, we expect regulators to be mindful that we, as humans, have certain biological needs and that there should be no encouragement for technologies that are dangerous in that they compromise the conditions for our very existence; secondly, given that we have a (self-interested) sense of which technological developments we would regard as beneficial, we will press regulators to support and prioritise such developments and, conversely, to reject developments that we judge to be contrary to our self-interest; and, thirdly, even where proposed technological developments are neither dangerous nor lacking utility, some will argue that they should be prohibited (or, at least, not encouraged)[80] because their development would be immoral.[81]

If we build on this analysis, we will argue that the paramount responsibility for regulators (whether they otherwise think like coherentists, regulatory-instrumentalists, or technocrats) is to protect, preserve, and promote:

- the essential conditions for human existence (given human biological needs);
- the generic conditions for human agency and self-development; and,
- the essential conditions for the development and practice of moral agency.

These, it bears repeating, are imperatives for regulators in all regulatory spaces, whether international or national, public or private. Of course, determining the nature of these conditions will not be a mechanical process and I do not assume that it will be without its points of controversy.[82] Nevertheless, let me give an indication of how I would understand the distinctive contribution of each segment of the commons.

78 Compare Roger Brownsword, 'Responsible Regulation: Prudence, Precaution and Stewardship' (2011) 62 *Northern Ireland Legal Quarterly* 573.

79 Compare, Stephen J. Morse, 'Uncontrollable Urges and Irrational People' (2002) 88 *Virginia Law Review* 1025, 1065-66.

80 Compare Roger Brownsword, 'Regulatory Coherence—A European Challenge' in Kai Purnhagen and Peter Rott (eds), *Varieties of European Economic Law and Regulation: Essays in Honour of Hans Micklitz* (Springer, 2014) 235, for discussion of the CJEU's decision and reasoning in Case C-34/10, Oliver Brüstle v Greenpeace e.V. (Grand Chamber, 18 October 2011).

81 Recall, e.g., Francis Fukuyama, *Our Posthuman Future* (London: Profile Books, 2002) for the argument that the development and application of modern biotechnologies, especially concerning human genetics, should not be permitted to compromise human dignity.

82 Moreover, even if it is agreed where the bottom lines are to be drawn, a community still has to decide how to handle proposals for uses of technologies that do not present a threat to any of the bottom line conditions.

In the first instance, regulators should take steps to protect, preserve and promote the natural ecosystem for human life.[83] At minimum, this entails that the physical well-being of humans must be secured; humans need oxygen, they need food and water, they need shelter, they need protection against contagious diseases, if they are sick they need whatever medical treatment is available, and they need to be protected against assaults by other humans or non-human beings. It follows that the intentional violation of such conditions is a crime against, not just the individual humans who are directly affected, but humanity itself.[84]

Secondly, the conditions for meaningful self-development and agency need to be constructed: there needs to be a sufficient sense of self and of self-esteem, as well as sufficient trust and confidence in one's fellow agents, together with sufficient predictability to plan, so as to operate in a way that is interactive and purposeful rather than merely defensive. Let me suggest that the distinctive capacities of prospective agents include being able:

- to freely choose one's own ends, goals, purposes and so on ('to do one's own thing')
- to understand instrumental reason
- to prescribe rules (for oneself and for others) and to be guided by rules (set by oneself or by others)
- to form a sense of one's own identity ('to be one's own person').

Accordingly, the essential conditions are those that support the exercise of these capacities.[85] With existence secured, and under the right conditions, human life becomes an opportunity for agents to be who they want to be, to have the projects that they want to have, to form the relationships that they want, to pursue the interests that they choose to have and so on. In the twenty-first century, no other view of human potential and aspiration is plausible; in the twenty-first century, it is axiomatic that humans are prospective agents and that agents need to be free.

The gist of these agency conditions is nicely expressed in a paper from the Royal Society and British Academy where, in a discussion of data governance and privacy, we read that:

Future concerns will likely relate to the freedom and capacity to create conditions in which we can flourish as individuals; governance will determine the social, political, legal and moral infrastructure that gives each person a sphere of protection through which they can explore who they are, with whom they want to relate and how they want to understand themselves, free from intrusion or limitation of choice.[86]

In this light, we can readily appreciate that unlike, say, Margaret Atwood's post-apocalyptic dystopia, *Oryx and Crake*[87] what is dystopian about George Orwell's *1984*[88] and Aldous Huxley's *Brave New World* [89] is not that human *existence* is compromised but that human *agency* is compromised.[90] We can appreciate, too, that today's dataveillance practices, as much as 1984's surveillance, 'may be doing less to deter destructive acts than [slowly to narrow] the range of tolerable thought and behaviour.'[91]

Thirdly, the commons must secure the conditions for an aspirant moral community, whether the particular community is guided by teleological or deontological standards, by rights or by duties, by communitarian or liberal or libertarian values, by virtue ethics, and so on. The generic context for moral community is impartial between competing moral visions, values, and ideals; but it must be conducive to 'moral' development and 'moral' agency in a formal sense. So, for example, in her discussion of techno-moral virtues, (sous) surveillance, and moral nudges, Shannon Vallor is rightly concerned that any employment of digital technologies to foster prosocial behaviour should respect the importance of conduct remaining 'our *own conscious activity and achievement* rather than passive, unthinking submission.'[92] She then invites readers to join her in imagining that Aristotle's Athens had been ruled by laws that 'operated in such an unobtrusive and frictionless manner that the citizens largely remained unaware of their content, their aims, or even their specific behavioral effects.'[93] In this regulatory environment, we are asked to imagine that Athenians 'almost never erred in moral life, either in individual or collective action.'[94] However, while these fictional Athenians are reliably prosocial, 'they cannot begin to explain *why* they act in good ways, why the ways they act *are* good, or *what* the good life for a human being or community might be.'[95] Without answers to these questions, we cannot treat these model citizens as moral beings. Quite simply, their moral agency is compromised by technologies (in this instance, legal rules) that do too much regulatory work.

Agents who reason impartially will understand that each human agent is a stakeholder in the commons where this represents the essential conditions for human existence together with the generic conditions of both self-regarding and other-regarding agency; and, it will be understood that these conditions must, therefore, be respected. While respect for the commons' conditions is binding on all human agents, it should be emphasised that these conditions do not rule out the possibility of prudential or moral pluralism. Rather, the commons represents the pre-conditions for both individual self-development and community debate, giving each agent the opportunity to develop his or her own view of what is prudent as well as what should be

83 Compare, J. Rockström et al, 'Planetary Boundaries: Exploring the Safe Operating Space for Humanity' (2009) 14 *Ecology and Society* 32 (http://www.ecologyandsociety.org/vol14/iss2/art32/) (last accessed November 14, 2016); and, Kate Raworth, *Doughnut Economics* (London: Random House Business Books, 2017) 43-53.

84 Compare Roger Brownsword, 'Crimes Against Humanity, Simple Crime, and Human Dignity' in Britta van Beers, Luigi Corrias, and Wouter Werner (eds), *Humanity across International Law and Biolaw* (Cambridge: Cambridge University Press, 2014) 87; and the Nuffield Council on Bioethics, Genome editing and human reproduction: social and ethical issues (London, July 2018), paras 3.72-3.78, for discussion of the interests of humanity (reaching beyond individual and social interests) and, in particular, of 'transgenerationalism'.

85 Compare the insightful analysis of the importance of such conditions in Maria Brincker, 'Privacy in Public and the Contextual Conditions of Agency' in Tjerk Timan, Bryce Clayton Newell, and Bert-Jaap Koops (eds), *Privacy in Public Space* (Cheltenham: Edward Elgar, 2017) 64; and, similarly, see Margaret Hu, 'Orwell's 1984 and a Fourth Amendment Cybersurveillance Nonintrusion Test' (2017) 92 *Washington Law Review* 1819, 1903-1904.

86 The Royal Society and British Academy, Connecting Debates on the Governance of Data and its Uses (London, December 2016) 5.

87 (London: Bloomsbury, 2003).

88 (London: Penguin Books, 1954) (first published 1949).

89 (London: Vintage Books, 2007) (first published 1932).

90 To be sure, there might be some doubt about whether the regulation of particular acts should be treated as a matter of the existence conditions or the agency conditions. For present purposes, however, resolving such a doubt is not a high priority. The important question is whether we are dealing with a bottom-line condition.

91 Frank Pasquale, *The Black Box Society* (Harvard University Press, 2015) 52.

92 Shannon Vallor, *Technology and the Virtues* (New York: Oxford University Press, 2016) 203 (emphasis in original).

93 Vallor (n 92).

94 Vallor (n 92).

95 Vallor (n 92) (emphasis in the original).

morally prohibited, permitted, or required. However, the articulation and contestation of both individual and collective perspectives (like all other human social acts, activities and practices) are predicated on the existence of the commons.

The regulatory responsibility to respect the community's fundamental values

Beyond the fundamental stewardship responsibilities, regulators are also responsible for ensuring that the fundamental values of their particular community are respected. Just as each individual human agent has the capacity to develop their own distinctive identity, the same is true if we scale this up to communities of human agents. There are common needs and interests but also distinctive identities.

From the middle of the Twentieth Century, many nation states have expressed their fundamental (constitutional) values in terms of respect for human rights and human dignity.[96] These values clearly intersect with the commons conditions and there is much to debate about the nature of this relationship and the extent of any overlap for example, if we understand the root idea of human dignity in terms of humans having the capacity freely to do the right thing for the right reason,[97] then human dignity reaches directly to the commons' conditions for moral agency.[98] However, those nation states that articulate their particular identities by the way in which they interpret their commitment to respect for human dignity are far from homogeneous. Whereas, in some communities, the emphasis of human dignity is on individual empowerment and autonomy, in others it is on constraints relating to the sanctity, non-commercialisation, non-commodification, and non-instrumentalisation of human life.[99] These differences in emphasis mean that communities articulate in very different ways on a range of beginning of life and end of life questions as well as questions of human enhancement, and so on.

Recalling the second wave of technological disruption, one question that should now be addressed is whether, and if so how far, a community sees itself as distinguished by its commitment to regulation by rule and by human agents. Is it distinctively East coast or West coast in its regulatory culture? In some smaller scale communities or self-regulating groups, there might be resistance to a technocratic approach because compliance that is guaranteed by technological means compromises the context for trust this might be the position, for example, in some business communities (where self-enforcing transactional technologies, such as blockchain, are rejected).[100] Or, again, a community might prefer to stick with regulation by rules and by human agents because, valuing public participation in setting standards as well as some flexibility in their application, it is worried that, with a more technocratic approach, there might be both reduced participation and a loss of flexibility.

If a community decides that it is generally happy with an approach that relies on technological features rather than rules, it then has to decide whether it is also happy for humans to be out of the loop. Where the technologies involve AI, the 'computer loop' might be the only loop that there is. As Shawn Bayern and his co-authors note, this raises an urgent question, namely: 'do we need to define essential tasks of the state that must be fulfilled by human beings under all circumstances?'[101] Furthermore, once a community is asking itself such questions, it will need to clarify its understanding of the relationship between humans and robots in particular, whether it treats robots as having moral status, or legal personality, and the like.[102]

In Europe, the latter question is still under relatively early discussion with a number of views being expressed.[103] However, in relation to the former question, Article 22 of the GDPR stakes out a default prohibition on solely automated decisions which have legal or other significant effects in relation to an individual and it then provides for humans to be brought back into the loop where the default does not apply. That said, the Article, as drafted, gives rise to many nice points of legal interpretation[104] and, more importantly, makes bold assumptions about the visibility and discrete nature of 'decisions' in technological infrastructures as well as about the confidence of (and in) human arbitrators who are brought back into the loop.[105]

It is, of course, essential that the fundamental values to which a particular community commits itself are consistent with (or cohere with) the commons conditions; and, if we are to talk about a new form of coherentism as I will suggest we should it should be focused in the first instance on ensuring that regulatory operations are so consistent.

The regulatory responsibility to seek an acceptable balance of interests

This takes us to the third tier of regulatory responsibility. As we have said, with the development of a regulatory-instrumentalist mind-set, we find that much of traditional tort and contract law is overtaken by an approach that seems to promote general policy objectives (such as supporting and encouraging beneficial innovation) while balancing this with countervailing interests. Given that different balances will appeal to different interest groups, finding an acceptable balance is a major challenge for regulators.[106]

96 See Roger Brownsword, 'Human Dignity from a Legal Perspective' in M.Duwell, J. Braavig, R. Brownsword, and D. Mieth (eds), *Cambridge Handbook of Human Dignity* (Cambridge: Cambridge University Press, 2014) 1.

97 For such a view, see Roger Brownsword, 'Human Dignity, Human Rights, and Simply Trying to Do the Right Thing' in Christopher McCrudden (ed), *Understanding Human Dignity* (Proceedings of the British Academy 192) (Oxford: The British Academy and Oxford University Press, 2013) 345; and, Developing a Modern Understanding of Human Dignity' in Dieter Grimm, Alexandra Kemmerer, and Christoph Möllers (eds), *Human Dignity in Context* (Baden-Baden: Nomos; Oxford: Hart, 2018) 299.

98 See, Roger Brownsword, 'From Erewhon to Alpha Go: For the Sake of Human Dignity Should We Destroy the Machines?' (2017) 9 *Law, Innovation and Technology* 117.

99 See Deryck Beyleveld and Roger Brownsword, *Human Dignity in Bioethics and Biolaw* (Oxford: Oxford University Press, 2001); Tim Caulfield and Roger Brownsword, 'Human Dignity: A Guide to Policy Making in the Biotechnology Era' (2006) 7 *Nature Reviews Genetics* 72; and Roger Brownsword, *Rights, Regulation and the Technological Revolution* (Oxford: Oxford University Press, 2008).

100 See, the excellent discussion in Karen E.C. Levy, 'Book-Smart, Not Street-Smart: Blockchain-Based Smart Contracts and The Social Workings of Law' (2017) 3 *Engaging Science, Technology, and Society* 1.

101 Shawn Bayern et al (n 36) 156.

102 See, e.g., Bert-Jaap Koops, Mireille Hildebrandt, and David-Olivier Jaquet-Chiffelle, 'Bridging the Accountability Gap: Rights for New Entities in the Information Society?' (2010) 11 *Minnesota Journal of Law, Science and Technology* 497; and Joanna J. Bryson, Mihailis E. Diamantis, and Thomas D. Grant, 'Of, for, and by the people: the legal lacuna of synthetic persons' (2017) 25 *Artif Intell Law* 273.

103 See, e.g., Thomas Burri, 'The EU is right to refuse legal personality for Artificial Intelligence' (opinion piece available at https://www.euractiv.com/section/digital/opinion/the-eu-is-right-to-refuse-legal-personality-for-artificial-intelligence/) (last accessed December 14, 2018).

104 See, e.g., Sandra Wachter, Brent Mittelstadt, and Luciano Floridi, 'Why a Right to Explanation of Automated Decision-Making Does Not Exist in the General Data Protection Regulation' (2017) 7 *International Data Privacy Law* 76.

105 On the latter point, see the fine analysis in Hin-Yan Liu, 'The Power Structure of Artificial Intelligence' (2018) 10 *Law, Innovation and Technology* 197, 222.

106 For a very helpful analysis of 'output' legitimacy (acceptability), see Chris Reed, 'Why Judges Need Jurisprudence in Cyberspace' (2018) 38 *Legal Studies* 263; and see, too, Roger Brownsword, 'Law, Regulation, and Tech-

Today, we have the perfect example of this challenge in the anxious debate about the responsibilities of Internet intermediaries, the argument being that they should be required to be far more active in monitoring the content they carry, failing which they should be held accountable for the negative consequences that ensue, where these consequences range from teenagers self-harming and committing suicide to parents declining vaccines for their children to acts of terrorism.[107] At the core of this debate is the question of whether intermediaries should be required to monitor content or simply act after the event by taking down offending content. In principle, we might argue that such intermediaries should be held strictly liable for any or some classes of illegal content; or that they should be liable if they fail to take reasonable care; or that they should be immunised against liability even though the content is illegal. If we take a position at the strict liability end of the range, we might worry that the liability regime is too burdensome to intermediaries and that online services will not expand in the way that we hope; but, if we take a position at the immunity end of the range, we might worry that this treats the Internet as an exception to the Rule of Law and that it becomes a hostage to fortune (inviting the illegal activities of copyright infringers, paedophiles, terrorists and so on). In practice, most legal systems balance these interests by taking a position that confers an immunity but only so long as the intermediaries do not have knowledge or notice of the illegal content. Predictably, now that the leading intermediaries are large US corporations with deep pockets, and not fledgling start-ups, many think that the time is ripe for the balance to be reviewed.[108] However, finding a balance that is generally acceptable, in both principle and practice, is another matter.

5.2 The Rule of Law

Technological management appeals because it promises to be more effective than rules; but, its brute instrumentalism demands that its use is conditioned by principles that give it legitimacy—otherwise, there is no reason why regulatees should at least acquiesce in its use. Although, as specified, technological management is materially different to the traditional legal enterprise of subjecting human conduct to the governance of *rules*, it is imperative that we apply the spirit of the Rule of Law to the regulatory use of technological measures.

Even though there are many conceptions of the Rule of Law, I take it that the spirit of this ideal is that it sets it face against both arbitrary governance and irresponsible citizenship.[109] Advocates of particular conceptions of the ideal will specify their own favoured set of conditions (procedural and substantive, thin or thick) for the Rule of Law which, in turn, will shape how we interpret the line between arbitrary and non-arbitrary governance as well as whether we judge citizens to be acting responsibly or irresponsibly in their response to acts of governance.[110] Viewed in this way, the Rule of Law represents a compact between, on the one hand, lawmakers, law-enforcers, law-interpreters, and law-appliers and, on the other hand, the citizenry. The understanding is that the actions of those who are in the position of the former should always be in accordance with the authorising constitutive rules (with whatever procedural and substantive conditions are specified); and that, provided that the relevant actions are in accordance with the constitutive rules, then citizens (including lawmakers, law-enforcers, law-interpreters, and law-appliers in their capacity as citizens) should respect the legal rules and decisions so made. In this sense, no one—whether acting offline or online—is above the law[111]; and the Rule of Law signifies that the law rules.

Similarly, if we apply this ideal to the acts of regulators—whether these are acts that set standards, or that monitor compliance, or that take corrective steps in response to non-compliance—then those acts should respect the constitutive limits and, in turn, they should be respected by regulatees provided that the constitutive rules are observed.[112]

In principle, we might also—and, indeed, I firmly believe that we should—apply the ideal of the Rule of Law to technological management.[113] The fact that regulators who employ technological management resort to a non-normative instrument does not mean that the compact is no longer relevant. On the one side, it remains important that the exercise of power through technological management is properly authorised and limited; and, on the other, although citizens might have less opportunity for 'non-compliance', it is important that the constraints imposed by technological management are respected. To be sure, the context of regulation by technological management is very different to that of a normative legal environment but the spirit and intent of the compact remains relevant.

The importance of the Rule of Law in an era of technological management should not be understated. Indeed, if we are to re-invent law for our technological times, one of the first priorities is to shake off the idea that brute force and coercive rules are the most dangerous expressions of regulatory power; the regulatory power to limit our practical options might be much less obvious but no less dangerous. Power, as Steven Lukes rightly says, 'is at its most effective when least observable.'[114]

While I cannot here specify a model Rule of Law for future communities, I suggest that the following conditions, reflecting the three-tiered scheme of regulatory responsibilities, merit serious consideration.[115]

nology: Supporting Innovation, Managing Risk and Respecting Values' in Todd Pittinsky (ed), *Handbook of Science, Technology and Society* (Cambridge: Cambridge University Press, 2019).

107 For a couple of recent examples, see Mark Bridge, Tom Knowles, and Kate Devlin, 'Footage of massacre spread over the Internet' *The Times*, March 16, 2019, p. 8 (reporting that, following the two mass shootings of Muslim worshippers in Christchurch, social media companies were being accused of 'aiding and abetting' terrorism), and Chris Smyth, 'Anti-vaccine posts could be banned' *The Times*, March 27, p. 2. For general discussion of the issues, see Lilian Edwards, '"With Great Power Comes Great Responsibility?": The Rise of Platform Liability' in Lilian Edwards (ed), *Law, Policy and the Internet* (Oxford: Hart, 2019) 253, esp 285-289. It should be noted that in the UK government's recent White Paper on this topic (n 9), it is proposed that there should be a new regulatory framework that 'will set clear standards to help companies ensure safety of users while protecting freedom of expression' (para 14).

108 For a particularly compelling analysis, see Marcelo Thompson, 'Beyond Gatekeeping: the Normative Responsibility of Internet Intermediaries' (2016) 18 *Vanderbilt Journal of Entertainment and Technology Law* 783; and Reed (n 107).

109 Compare Deryck Beyleveld and Roger Brownsword, *Law as a Moral Judgment* (London: Sweet and Maxwell, 1986; Sheffield: Sheffield Academic Press, 1994) Ch. 9; and Roger Brownsword, *Law, Technology and Society: Re-imagining the Regulatory Environment* (Abingdon: Routledge, 2019) Ch. 5.

110 Generally, see Joseph Raz, 'The Rule of Law and its Virtues' (1977) 93 *LQR* 195; and David Dyzenhaus, 'Recrafting the Rule of Law' in David Dyzenhaus (ed), *Recrafting the Rule of Law* (Oxford: Hart, 1999) 1.

111 Compare Joel R. Reidenberg, 'Technology and Internet Jurisdiction' (2005) 153 *University of Pennsylvania Law Review* 1951, resisting the claims of the 'Internet separatists' and defending the application of the Rule of Law to online environments.

112 Compare Karen Yeung, *Securing Compliance* (Oxford: Hart, 2004).

113 Compare Mireille Hildebrandt, *Smart Technologies and the End(s) of Law* (Cheltenham: Edward Elgar, 2015).

114 Steven Lukes, *Power: A Radical View* (2nd ed) (Basingstoke: Palgrave Macmillan, 2005) 1.

115 Compare Roger Brownsword, 'The Rule of Law, Rules of Law, and Techno-

First, for any community, it is imperative that technological management (just as with rules and standards) does not compromise the essential conditions for human social existence (the commons). The Rule of Law should open by emphasising that the protection and maintenance of the commons is always the primary responsibility of regulators. Moreover, all uses of technological management, whether by public regulators or by private regulators or actors should respect this fundamental responsibility.

Secondly, where the aspiration is not simply to be a moral community (a community committed to the primacy of moral reason) but a particular kind of moral community, then it will be a condition of the Rule of Law that the use of technological management (just as with rules and standards) should be consistent with its particular constitutive features whether those features are, for instance, liberal or communitarian in nature, rights-based or utilitarian, and so on. Such is the logic of the second tier of responsibility.

As we have said, many modern communities have articulated their constitutive values in terms of respect for human rights and human dignity.[116] In an age of technological management, this might translate into a human right (or corresponding duties derived from respect for human dignity) to know whether one is interacting or transacting with a robot, to being cared for by humans (rather than robots which can appear to care but without really caring),[117] to having a right to have 'bad news' conveyed by another human,[118] and to reserving the possibility of an appeal to a human arbitrator against a decision that triggers an application of technological management that forces or precludes a particular act or that excludes a particular person or class of persons.[119]

Looking ahead, one (possibly counter-intuitive) thought is that a community might attach particular value (based on its interpretation of respect for human rights and human dignity) to preserving both human officials (rather than machines) and rules (rather than technological measures) in the core areas of the criminal justice system.[120]

To ring-fence core crime in this way promises to retain some flexibility in the application of rules that carry serious penalties for their infringement as well as preserving an important zone for moral development (and display of moral virtue). Indeed, in some communities, this zone might be thought to be so critical to the very possibility of moral development that the eschewal of technological solutions is seen as reaching back to the commons conditions themselves.[121]

Thirdly, where the use of technological management is proposed as part of a risk management package, so long as the community is committed to the ideals of deliberative democracy, it will be a condition of the Rule of Law that there needs to be a transparent and inclusive public debate about the terms of the package. It will be a condition that all views should be heard with regard to whether the package amounts to both an acceptable balance of benefit and risk as well as representing a fair distribution of such risk and benefit (including adequate compensatory provisions). Before the particular package can command respect, it needs to be somewhere on the spectrum of reasonableness. This is not to suggest that all regulatees must agree that the package is optimal; but it must at least be reasonable in the weak sense that it is not a package that is so unreasonable that no rational regulator could, in good faith, adopt it. Such is the shape of the third tier of responsibility.

For example, where technologically managed places or products operate dynamically, making decisions case-by-case or situation-by-situation, then one of the outcomes of the public debate might be that the possibility of a human override is reserved. In the case of driverless cars, for instance, we might want to give agents the opportunity to take control of the vehicle in order to deal with some hard moral choice (whether of a 'trolley' or a 'tunnel' nature) or to respond to an emergency (perhaps involving a 'rescue' of some kind).[122]

Similarly, there might be a condition that interventions involving technological management should be reversible—a condition that might be particularly important if measures of this kind are designed not only into products and places but also into people, as might be the case if regulators contemplate making interventions in not only the coding of product software but also the genomic coding of particular individuals. It should be noted, however, that while reversibility might speak to the acceptability of a technological measure, it might go deeper, to either second or first tier responsibilities.

Fourthly, where following community debate or public deliberation, particular limits on the use of technological management have been agreed, those limits should be respected. Clearly, it would be an abuse of power to exceed such limits. In this sense, the use of technological management should be congruent with the particular rules agreed for its use, as well as being coherent with the community's constitutive rules.[123]

Fifthly, the community will want to be satisfied that the use of technological measures is accompanied by proper mechanisms for accountability. When there are problems, or when things go wrong, there need to be clear, accessible, and intelligible lines of accountability. It needs to be clear who is to be held to account as well as how they are to be held to account; and, the accounting itself must be meaningful.[124]

logical Management' *Amsterdam Law School Research Paper* No. 2017-35 (2017) 9-17. Available at: https://ssrn.com/abstract=3005914. See, too, Gavaghan (n 33), where, at 135, it is suggested that, in addition to asking the general question about whether a measure is 'likely to be effective, what we think of the values it embodies, whether the likely benefit is worth the cost, and so forth', we should ask whether technological measures are (i) visible, (ii) flexible, (iii) simply enforcing rules already agreed upon by democratic means, and (iv) employing unusually intrusive or inflexible means of enforcement.

116 See, Roger Brownsword, 'Human Dignity from a Legal Perspective' in Marcus Duwell, Jens Braavig, Roger Brownsword, and Dietmar Mieth (eds), *Cambridge Handbook of Human Dignity* (Cambridge: Cambridge University Press, 2014) 1.

117 See, e.g., Sherry Turkle, *Alone Together* (New York: Basic Books, 2011) esp. at 281-282 (concerning the case of Richard).

118 Earlier this year, it was reported that Ernest Quintana's family were shocked when they saw that a 'robot' displaying a doctor on a screen was used to tell Ernest that doctors (at the Californian hospital where he was a patient) could do no more for him and that he would die soon: see Michael Cook, 'Bedside manner 101: How to deliver very bad news' Bioedge (March 18, 2019). Available at https://www.bioedge.org/bioethics/bedside-manner-101-how-to-deliver-very-bad-news/12998 (last accessed April 3, 2019).

119 Compare Gavaghan (n 33). However, the extent to which the possibility of human intervention can make much practical difference when smart machines are known to outperform humans is moot. For insightful discussion, see Hin-Yan Liu (n 105).

120 Compare Roger Brownsword and Alon Harel, 'Law, Liberty and Technology Criminal Justice in the Context of Smart Machines' (2019) 12 *International Journal of Law in Context* (forthcoming) and Deryck Beyleveld and Roger Brownsword, 'Punitive and Preventive Justice in an Era of Profiling, Smart Prediction and Practical Preclusion' (2019) 12 *International Journal of Law in Context* (forthcoming).

121 Compare the discussion in Roger Brownsword (n 98).

122 For discussion of such moral hard choices, see Roger Brownsword, *Law, Technology and Society Re-imagining the Regulatory Environment* (Abingdon: Routledge, 2019) 249-251.

123 Compare Gavaghan (n 33).

124 See Joshua A. Kroll, Joanna Huey, Solon Barocas, Edward W. Felten, Joel R.

Sixthly, a community might be concerned that the use of technological management will encourage some mission creep. If so, it might stipulate that the restrictive scope of measures of technological management or their forcing range should be no greater than would be the case were a rule to be used for the particular purpose. In this sense, the restrictive sweep of technological management should be, at most, co-extensive with that of the equivalent (shadow) rule.[125]

Seventhly, it is implicit in the Fullerian principles of legality[126] that regulators should not try to trick or trap regulatees; and this is a principle that is applicable whether it is rules or technological measures that are employed as regulatory instruments. Accordingly, it should be a condition of the Rule of Law that technological management should not be used in ways that trick or trap regulatees and that, in this sense, the administration of a regime of technological management should be in line with the reasonable expectations of regulatees (implying that regulatees should be put on notice that technological management is in operation).[127] Crucially, if the default position in a technologically managed regulatory environment is that, where an act is found to be available, it should be treated as permissible, then regulatees should not be penalised for doing the act on the good faith basis that, because it is available, it is a lawful option.

Eighthly, regulatees might also expect there to be a measure of public scrutiny of the private use of technological management. Even if public regulators respect the conditions set by regulatees, it will not suffice if private regulators are left free to use technological management in ways that compromise the planetary conditions or the essential context for agency, or violate the community's constitutive principles, or exceed the agreed and authorised limits for its use. Accordingly, it should be a condition of the Rule of Law that the *private* use of technological management should be compatible with the general principles for its use.

5.3 A New Coherentism

In the bigger picture of regulatory responsibilities, where the paramount responsibility is to ensure that no harm is done to the commons, we might wonder whether a traditional coherentist mind-set is appropriate. If regulators think in such a coherentist way, they might fail to take the necessary protective steps steps that might involve new rules, or the use of measures of technological management, or both. While the commons is being compromised, we might fear, coherentists will be concerned only with the integrity of doctrine.

Such a concern invites the thought that a regulatory-instrumentalist approach is a better default but it is only so if regulators are focused on the relevant risks namely, the risks presented by technological development to the commons' conditions. Moreover, we might want to add that regulatory-instrumentalism with this particular risk focus is only a better default if it is applied with a suitably precautionary mentality. Regulators need to understand that compromising the commons is always the worst-case scenario.[128] Alongside such a default, a technocratic approach might well be appropriate. For example, if we believe that a rule-based approach cannot protect the planetary boundaries, then a geo-engineering approach might be the answer.[129] However, it needs to be borne in mind that, with a resort to technological management, there is potentially more than one kind of risk to the commons: an ineffective attempt to manage risks to the existence conditions might actually make things worse; and an effective intervention for the sake of the existence conditions might compromise the conditions for self-development and moral agency (because both autonomy and virtue presuppose a context in which one acts freely).

Accordingly, the third element in the re-invention of law is the articulation of a 'new coherentism'. New coherentism reminds all those who act as regulators of two things: first, that their most urgent regulatory focus should be on the commons' conditions; and, secondly, that, whatever their interventions, and particularly where they take a technocratic approach, their acts must always be compatible with the preservation of the commons.

In future, the Courts albeit the locus for traditional coherentist thinking will have an important role to play in bringing new coherentism to bear on the use of technological measures. Most importantly, it will be for the Courts to review the legality of any measure that is challenged relative to the authorising and constitutive rules; and, above all, to check that particular instances of technological management are consistent with the commons-protecting ideals that are inscribed in the Rule of Law. In short, although traditional coherentism might have been prized by private lawyers, the new coherentism is material to questions of public and constitutional law, and beyond that it reaches through to the maintenance of the essential conditions for any community of human agents. Moreover, whatever the significance of the contested distinction between the public and the private, it is certainly not that private regulators have a licence to pursue their own interests regardless of their responsibilties for the preservation and protection of the commons.

With a new coherentist mind-set, it is not a matter of checking for internal doctrinal consistency, nor checking that a measure is fit for its particular regulatory purpose. Rather, a renewed ideal of coherence should start with the paramount responsibility of regulators, namely, the protection and preservation of the commons. All regulatory interventions should cohere with that responsibility. This means that the conditions for both human existence and the context for flourishing agency should be respected. In line with such thinking, in 2017, when researchers met at Asilomar in California to develop a set of precautionary guidelines for the use of AI, it was agreed (in Principle 21) that 'risks posed by AI systems, especially catastrophic or existential risks, must be subject to planning and mitigation efforts commensurate with their expected impact.'[130]

Moreover, as we have emphasised, if the commons is to be respected, technological management should not be employed in ways that compromise the context for agency and moral community.

Reidenberg, David G. Robinson, and Harlan Yu, 'Accountable Algorithms' (2017) 165 *University of Pennsylvania Law Review* 633, 702-704.

125 Compare Gavaghan (n 33).

126 Seminally, see Lon L. Fuller, *The Morality of Law* (New Haven: Yale University Press, 1969). For an application of the Fullerian principles to particular instances of cyberlaw, see Chris Reed, 'How to Make Bad Law: Lessons from Cyberspace' (2010) 73 *MLR* 903, esp at 914-916.

127 Compare Gavaghan (n 33) on visibility, at 135-137 (do we know that technological measures are employed, do we know that they are in operation in a particular place or at a particular time, and do we know the precise details or limits of such measures?).

128 Compare Deryck Beyleveld and Roger Brownsword, 'Complex Technology, Complex Calculations: Uses and Abuses of Precautionary Reasoning in Law' in Marcus Duwell and Paul Sollie (eds), *Evaluating New Technologies: Methodological Problems for the Ethical Assessment of Technological Developments* (Dordrecht: Springer, 2009) 175; and 'Emerging Technologies, Extreme Uncertainty, and the Principle of Rational Precautionary Reasoning (2012) 4 *Law Innovation and Technology* 35.

129 For discussion, see Jesse Reynolds, 'Solar Climate Engineering, Law, and Regulation' in Brownsword, Scotford, and Yeung (n 26) 799.

130 Available at https://futureoflife.org/ai-principles/ (last accessed March 18, 2019)

Consider, for example, the much debated and protean concept of privacy. A popular view is that respect for privacy should be applied in a 'contextual' way.[131] However, there is Context and there are contexts. There is Context (in the sense of the commons) and then there are many contexts that rely on the integrity of the commons. So, if it is judged that privacy reaches through to the interests that agents necessarily have in the commons' conditions, particularly in the conditions for self-development and agency, it is neither rational nor reasonable for agents, individually or collectively, to authorise acts that compromise these conditions (unless they do so in order to protect some more important condition of the commons). As Maria Brincker expresses this point:

> Agents act in relation not to singular affordances but to affordance spaces: choices are always situated calibrations of multiple interests and purposes given the perceived opportunities. To assess the values and risks of potential actions we need to have expectations regarding the consequences of those actions.[132]

It follows, argues Brincker, that without some degree of privacy 'our very ability to act as autonomous and purposive agents' might be compromised.[133] On the other hand, if privacy (and, likewise, data protection) is judged to be simply a legitimate informational interest that has to be weighed in an all things considered balance of interests, then we should recognise that what each community will recognise as a privacy interest and as an acceptable balance of interests might well change over time. To this extent, our reasonable expectations of privacy might be both 'contextual' and contingent on social practices.[134] That said, a community might wish to define itself by giving privacy an elevated status (as a right or a fundamental right) which regulators will then need to respect as an overriding interest. However, no community can rationally define itself in ways that are incompatible with the common interest in the essential infrastructural conditions.

Next, measures of technological management should cohere with the particular constitutive values of the community such as respect for human rights and human dignity, the way that non-human agents are to be treated, and so on and its particular articulation of the Rule of Law. At Asilomar, it was agreed (in Principle 11) that 'AI systems should be designed and operated so as to be compatible with ideals of human dignity, rights, freedoms, and cultural diversity'; and, in its recently published report, *Ethics Guidelines for Trustworthy AI*, the EC High-Level Group on Artificial Intelligence has explicitly based its guidance on the regional commitment to human rights.[135] As the Expert Group interprets this commitment, the foundational respect for human rights is 'rooted in respect for human dignity thereby reflecting what we describe as a 'human-centric' approach in which the human being enjoys a unique and inalienable moral status of primacy in civil, political, economic and social fields.'[136] Moreover, although the report reflects a broad spectrum of concerns, the Group recognises that its guidelines also have a dimension of depth. Accordingly, having cautioned that some trade-offs might have to be made, the Group then emphasises that certain 'fundamental rights and correlated principles are absolute and cannot be subject to a balancing exercise (e.g. human dignity).'[137] No doubt, the Courts will face many challenges in developing a coherent account of these principles (for example, with regard to the interpretation of 'humanity') but the critical point is that they should always be guided by a new coherentist understanding of their role and responsibility.

There will also be challenges to technological management on procedural grounds. Once again, there will be work for the Courts. Where explicit procedures are laid out for the adoption of technological management, the Courts will be involved in a familiar reviewing role. However, there might also be some doctrinal issues of coherence that arise—for example, where it is argued that the explicit procedural requirements have some further procedural entailments; or where the Courts, having developed their own implicit procedural laws (such as a practice raising a legitimate expectation of consultation), find that the body of doctrine is not internally coherent.

Coherence might be an ideal that is dear to the hearts of private lawyers but, in an era of technological management, it is once coherence is brought into the body of public law that we see its full regulatory significance. Regulation, whether normative or non-normative, will lack coherence if the procedures or purposes that accompany it are out of line with the authorising or constitutive rules that take us back to the Rule of Law itself; and, regulation will be fundamentally incoherent if it is out of line with the responsibility for maintaining the commons. In short, we can continue to treat coherence as an ideal that checks backwards, sideways, and upwards; but, the re-imagination of this ideal necessitates its engagement with both the full range of regulatory responsibilities and the full repertoire of regulatory instruments.

5.4 Institutional Design

If we are to be properly geared for the discharge of our regulatory responsibilities, this might call for some redesigning of the institutions on which we rely both nationally and internationally. While we can expect national regulators to deal with the routine balancing of interests within their communities as well as respecting the distinctive values of their particular community, the stewardship of the commons seems to call for international oversight. We can start with some remarks about the arrangements nationally for engaging with emerging technologies and then we can turn to the possible international regulation of the commons.

The design of national institutions

In the United Kingdom (and, I suspect, in many other nation states), there are two contrasting features in the institutional arrangements that we have for engaging with and regulating new technologies. On the one hand, there is no standard operating procedure for undertaking the initial review of such technologies; and, on the other hand, the Rule of Law in conjunction with democracy dictates that the Courts should settle disputes in accordance with established legal princi-

131 See, e.g., Daniel J. Solove, *Understanding Privacy* (Cambridge, Mass.: Harvard University Press, 2008), and Helen Nissenbaum, Privacy in Context (Stanford: Stanford University Press, 2010).

132 Maria Brincker (n 85) 88. Similarly, see Margaret Hu, 'Orwell's 1984 and a Fourth Amendment Cybersurveillance Nonintrusion Test' (2017) 92 *Washington Law Review* 1819, 1903-1904.

133 Brincker (n 85) 64.

134 Compare the insightful analysis in Bert-Jaap Koops and Ronald Leenes, '"Code" and the Slow Erosion of Privacy' (2005) 12 *Michigan Telecommunications and Technology Law Review* 115.

135 High-Level Expert Group on Artificial Intelligence, Ethics Guidelines for Trustworthy AI (European Commission, Brussels, April 8, 2019). Compare, too, the five overarching principles in the House of Lords Select Committee on Artificial Intelligence's Report on AI in the UK; ready, willing and able? (Report of Session 2017-19, published 16 April 2017, HL Paper 100) at para 417: available at https://publications.parliament.uk/pa/ld201719/ldselect/ldai/100/10007.htm#_idTextAnchor025 (last accessed August 11, 2018).; and the Google White Paper, Perspectives on issues in AI governance (January, 2019), see https://ai.google/static/documents/perspectives-on-issues-in-ai-governance.pdf.

136 High-Level Expert Group (n 135) 10.

137 High-Level Expert Group (n 135) 13.

ples and that it is for the Legislature and the Executive to formulate and agree public policies, plans and priorities. In other words, while there is no expectation about who will undertake the initial review or how that review will be approached, we have very definite expectations about the role and reasoning of judges and advocates in the Courts (where the discourse is coherentist) and similarly about the policy-making members of the Legislature and Executive (where the discourse is regulatory-instrumentalist). The question is: where in this institutional design do we find the responsibility for stewardship of the commons and for the community's distinctive values?

To start with the initial engagement with, and review of, an emerging technology, it seems to be largely a matter of happenstance as to who addresses the issue and how it is addressed at any rate, this is the case in the UK. To pick up an earlier example, in the late 1970s, when techniques for assisted conception were being developed and applied, but also being seriously questioned, the response of the UK government was to set up a Committee of Inquiry chaired by Mary Warnock. In 1984, the Committee's report (the Warnock Report) was published.[138] However, it was not until 1990, and after much debate in Parliament, that the framework legislation, the Human Fertilisation and Embryology Act 1990, was enacted. This process, taking the best part of a decade, is regularly held up as an example of best practice when dealing with emerging technologies. Nevertheless, this methodology is not in any sense the standard operating procedure for engaging with new technologies—indeed, there is no such procedure.

The fact of the matter is that legal and regulatory responses to emerging technologies vary from one technology to another, from one legal system to another, and from one time to another. Sometimes, there is extensive public engagement, sometimes not. On occasion, special Commissions (such as the now defunct Human Genetics Commission in the UK) have been set up with a dedicated oversight remit; and there have been examples of standing technology foresight commissions (such as the US Office of Technology Assessment)[139]; but, often, there is nothing of this kind. Most importantly, questions about new technologies sometimes surface, first, in litigation (leaving it to the Courts to determine how to respond) and, at other times, they are presented to the legislature (as was the case with assisted conception).

With regard to the question of which regulatory body engages with new technologies and how, there can of course be some local agency features that shape the answers. Where, as in the United States, there is a particular regulatory array with each agency having its own remit, a new technology might be considered in just one lead agency or it might be assessed in several agencies.[140] Once again, there is a degree of happenstance about this. Nevertheless, in a preliminary way, we can make three general points.

First, if the question (such as that posed by a compensatory claim made by a claimant who alleges harm caused by a new technology) is put to the Courts, their responsibility for the integrity of the law will push them towards a traditional coherentist assessment. Typically, courts are neither sufficiently resourced nor mandated to undertake a risk assessment let alone adopt a risk management strategy (unless the legislature has already put in place a scheme that delegates such a responsibility to the courts).[141]

Secondly, if the question finds its way into the legislative arena, it is much more likely that politicians will engage with it in a regulatory-instrumentalist way; and, once the possibility of technological measures gets onto the radar, it is much more likely that (as with institutions in the EU) we will see a more technocratic mind-set.

Thirdly, if leaving so much to chance seems unsatisfactory, then it is arguable that there needs to be a body that is charged with undertaking the preliminary engagement with new technologies. The remit and challenge for such a body would be to ensure that there is no harm to the commons; to try to channel such technologies to our most urgent needs (relative to the commons); and, to help each community to address the question of the kind of society that it distinctively wants to be—doing all that, moreover, in a context of rapid social and technological change. As Wendell Wallach rightly insists:

> Bowing to political and economic imperatives is not sufficient. Nor is it acceptable to defer to the mechanistic unfolding of technological possibilities. In a democratic society, we—the public—should give approval to the futures being created. At this critical juncture in history, an informed conversation must take place before we can properly give our assent or dissent.[142]

Granted, the notion that we can build agencies that are fully fit for such purposes might be an impossible dream. Nevertheless, I join those who argue that this is the right time to set up a suitably constituted body,[143] one that would underline our responsibilities for the commons as well as facilitating the development of each community's regulatory and social licence for these technologies.[144] Possibly this might be along the lines of the Centre for Data Ethics and Innovation as announced by the UK government in late 2017,[145] the wide-ranging terms of reference for which require it to analyse and anticipate risks and opportunities, to agree and articulate best practice, and to advise on the need for action. However, this is a matter for further discussion.

138 Report of the Committee of Inquiry into Human Fertilisation and Embryology (London: HMSO, Cm. 9314, 1984).

139 On which, see Bruce Bimber, *The Politics of Expertise in Congress* (Albany: State University of New York Press, 1996) charting the rise and fall of the Office and drawing out some important tensions between 'neutrality' and 'politicisation' in the work of such agencies.

140 Compare, Albert C. Lin, 'Size Matters: Regulating Nanotechnology' (2007) 31 *Harvard Environmental Law Review* 349.

141 Perhaps we should view Patent Offices in this light. In the 1980s, there were major decisions to be made about the patentability of biotechnological products and processes, models of which could not be brought into the Office to demonstrate how they worked and which also raised complex moral issues. For extended discussion, see Alain Pottage and Brad Sherman, *Figures of Invention: A History of Modern Patent Law* (Oxford: Oxford University Press, 2010); and, on the moral dimension of these debates, see Deryck Beyleveld and Roger Brownsword, *Mice, Morality and Patents* (London: Common Law Institute of Intellectual Property, 1993).

142 See, Wendell Wallach, *A Dangerous Master* (Basic Books, 2015) 10.

143 Amongst many matters in this paper that invite further discussion, the composition of such a Commission invites debate. See, too, Wallach (n 142) Chs 14-15.

144 Compare Geoff Mulgan's proposal for the establishment of a Machine Intelligence Commission: available at http://www.nesta.org.uk/blog/machine-intelligence-commission-uk (blog 'A machine intelligence commission for the UK', February 22, 2016: last accessed December 11, 2016); Olly Bustom et al, *An Intelligent Future? Maximising the Opportunities and Minimising the Risks of Artificial Intelligence in the UK* (Future Advocacy, London, October 2016) (proposing a Standing Commission on AI to examine the social, ethical, and legal implications of recent and potential developments in AI); HC Science and Technology Committee, Robotics and Artificial Intelligence HC 145 2016-17.

145 See 'Autumn Budget 2017: 25 things you need to know' (H.M. Treasury, November 22, 2017) point 16: available at https://www.gov.uk/government/news/autumn-budget-2017-25-things-you-need-to-know (last accessed November 25, 2017). Compare, too, discussion in Part 3 of the White Paper on online harms (n 9) with regard to whether functions and responsibilities of the proposed independent regulator should be undertaken by a new or by an existing regulatory body.

In the light of this, consider briefly the much-debated question of who should be liable for what if there are accidents that involve autonomous vehicles. It goes without saying that it makes little sense to try, in a coherentist way, to apply the principles for judging the negligence of human drivers to questions of liability concerning vehicles in which there is no human in control and where the nature of the technology militates against simple causal accounts when things 'go wrong'. Yet, if these questions are taken up in the courts, we must expect that judges (reasoning like coherentists) will try to apply notions of a reasonable standard of care, proximate cause, and so on, to determine responsibility for very complex technological failures.[146] Indeed, when Joshua Brown was killed while driving his Tesla S car in autopilot mode,[147] Tesla (presumably anticipating litigation or a discourse of fault and responsibility) were quick to highlight the safety record of their cars, to suggest that drivers of their cars needed to remain alert, and to deny that they themselves were careless in any way. By contrast, if regulators in a legislative setting approach the question of liability and compensation with a risk-management mind-set, they will not need to chase after questions of fault or, at any rate, as in the UK Automated and Electric Vehicles Act 2018, insurance and compensation will come first with insurers (and owners) of automated vehicles then able to pursue existing (fault-based) common law claims. In this way, the challenge will be to articulate the most acceptable (and financially workable) compensatory arrangements that accommodate the interest in transport innovation with the interest in the safety of passengers and pedestrians.[148] Ideally, regulators should take a view only after an independent emerging technologies body (of the kind that we do not, but surely should, have) has informed and stimulated public debate.

International stewardship of the commons

The commons is not confined to particular nation states. The conditions for human existence on planet Earth are relevant to all nation states and can be impacted by each nation state's activities. The same applies where nation states interfere with the conditions for flourishing agency beyond their own national borders. Whether in relation to the conditions for human existence or for the enjoyment of human agency, there can be cross-border spill-over effects. Accordingly, if the essential infrastructure for human social existence is to be secured, this implies that there needs to be a considerable degree of international co-ordination and shared responsibility.[149]

Given that the international regulatory architecture is already extensive, we might think that securing the commons will only require some minor adjustments or additions. On the other hand, stewardship of the kind that is contemplated requires a distinctive and dedicated approach. It might be, therefore, that we need to have bespoke international laws and new international agencies to take this project forward.[150] Moreover, because politics tends to operate with short-term horizons, it also implies that the regulatory stewards should have some independence from the political branch, but not of course that they should be exempt from the Rule of Law's culture of accountability and justification.[151]

Whatever the ideal design, we have to take into account the realities of international relations. One of these realities is that there are at least three kinds of international citizens: first, there are functioning states amongst whom many are good citizens of the international order (respecting the rules of international law); secondly, there are functioning states that are also superpowers (who largely dictate and veto international initiatives as well as playing by their own rules); and, thirdly, there are rogue states (who play by no rules).[152] If the regulatory stewards were drawn from the good citizens, that might be fine insofar as an agency so populated would be focused on the right question and motivated by concerns for the common interest of humans. However, they might find that they are blocked in their efforts to introduce necessary measures of technological management; and, without the support of others, they will be in no position to ensure compliance with whatever precautionary standards they might propose.

A second reality is that, where the missions of international agencies include a number of objectives, trade (rather than human rights or environmental concerns) will often be prioritised.[153] It follows that, if the regulatory stewards are within an international agency, the mission must be limited to the protection of the commons. Even then, there would be no guarantee that the stewards would be immunised against the usual risks of regulatory capture and corruption. In short, unless the culture of international relations is supportive of the stewards, even the ideal regulatory design is likely to fail.

The moral seems to be that, if the common interest is to be pursued, this is a battle for hearts and minds. As Neil Walker has remarked in relation to global law, our future prospects depend on 'our ability to persuade ourselves and each other of what we hold in common and of the value of holding that in common.'[154] An international agency with a mission to preserve the commons might make some progress in extending the pool of good citizens but to have any chance of success all nation states need to be on board.

6. Concluding Remarks

In this article, I have described two ways in which law is disrupted by new technologies. To some extent, this is an old story. From the industrial revolution onwards, legal rules have needed remedial attention as their deficiencies are exposed as it becomes apparent that the prevailing rules are not fit for regulatory purpose. That said, the very idea of *a rule not being fit for regulatory purpose* is itself expressive of a radically disrupted way of thinking. Crucially, though, this old story is now joined by a new disruptive story in which it is the taken-for-

146 I take it that, if autonomous vehicles have to be at least as safe as driven vehicles, there would be a difficulty in presenting them as 'dangerous' in a way that would get a product liability claim to first base.

147 Reported at https://www.theguardian.com/technology/2016/jun/30/tesla-autopilot-death-self-driving-car-elon-musk (last accessed November 14, 2017).

148 For analysis and proposals, see Maurice Schellekens, 'No fault compensation schemes for self-driving vehicles' (2018) 10 *Law, Innovation and Technology* 314.

149 See David A. Wirth, 'Engineering the Climate: Geoengineering as a Challenge to International Governance' (2013) 40 *Boston College Environmental Affairs Law Review* 413, esp. at 430-436.

150 Compare, e.g., Seth D. Baum and Grant S. Wilson, 'The Ethics of Global Catastrophic Risk from Dual Use Bioengineering' (2013) 4 *Ethics in Biology, Engineering and Medicine* 59; Grant Wilson, 'Minimizing global catastrophic and existential risks from emerging technologies through international law' (2013) 31 *Virginia Environmental Law Journal* 307; and Dennis Pamlin and Stuart Armstrong, 'Twelve risks that threaten human civilisation: The case for a new risk category' (Oxford: Global Challenges Foundation, 2015) 182 (mooting the possibility of establishing a Global Risk Organisation, initially only with monitoring powers).

151 See, too, Roger Brownsword (n 78).

152 Compare Gerry Simpson, *Great Powers and Outlaw States* (Cambridge: Cambridge University Press, 2009).

153 See, e.g., Sheldon Leader, 'Collateralism' in Roger Brownsword (ed) *Global Governance and the Quest for Justice Vol IV: Human Rights* (Oxford: Hart, 2004) 53.

154 Neil Walker, *Intimations of Global Law* (Cambridge: Cambridge University Press, 2015) 199.

granted assumption that social ordering is to be achieved through rules that is challenged and, concomitantly, that the Rule of Law is exclusively about rule by rules and about application and enforcement of the published rules by human agents. Regulation in future might be more a matter of a conversation between smart machines than a debate in a legislative forum where the participants are human agents.

Given such disruption, what should we do? I have suggested that we should reframe our thinking, re-imagining law as a part of a much more inclusive regulatory environment, an environment that features not only rule-based normative signals but also measures of non-normative technological management. So re-imagined, we can develop a jurisprudence that marks up the credentials of rules rather than technological measures, and vice versa.

There is no guarantee that rules and technological measures can peacefully co-exist. However, if we are to re-invent law, I have suggested that we first need to put in place a grounded and hierarchically ordered scheme of regulatory responsibilities. That scheme can then be used to inform each community's articulation of the Rule of Law (constraining and authorising the use of measures of technological management) and it can be taken forward through a new and revitalised form of coherentist thinking together with new institutional arrangements.

Rationally, humans should need little persuading: what we all have in common is a fundamental reliance on a critical infrastructure; if that infrastructure is compromised, the prospects for any kind of legal or regulatory activity, or any kind of persuasive or communicative activity, indeed for any kind of human social activity, are diminished. If we value anything, if we are positively disposed towards anything, we must value the commons. If we cannot agree on that, and if we cannot agree that the fundamental role of law is to ensure that power is exercised only in ways that are compatible with the preservation of the infrastructure of all other infrastructures, then the story of disruption, re-imagination and re-invention certainly will not end well.

Finally, I should emphasise that when I say 'we' I mean 'especially we lawyers'. Quite possibly, it will be those lawyers who have an interest in regulation or in emerging technologies who are in the vanguard. However, I would not want to limit responsibility in this way. For, if, as lawyers, we understand how this story should end, then we have a special responsibility to do our best to ensure that it does go well. In this story, we are not merely observers; we have a responsibility for constitutions and for codes, but above all we have a responsibility for the commons and for humanity.

03

Technology and Regulation

Artificial Intelligence: Some ethical issues and regulatory challenges

Mark Coeckelbergh*

responsibility, privacy, transparency, explainability, bias, AI policy, ethics by design

This article offers a brief overview of some of the ethical challenges raised by artificial intelligence (AI), in particular machine learning and data science, and summarizes and discusses a number of challenges for near-future regulation in this area. This includes the difficulties of moving from principles to more concrete measures and problems with implementing ethics by design and responsible innovation.

mark.coeckelbergh@univie.ac.at

1. Introduction

AI is already having a pervasive impact today as it is embedded in everyday digital technological systems, and its promises and attractions are likely to increase this impact in the near future. It is likely to have impact in many domains such as transport, marketing, health care, finance, security, science, education, entertainment, agriculture, and manufacturing.

While AI is likely to have many benefits, it also raises a number of ethical issues, some of which are well-known (e.g., privacy) and others which have to do with specific technologies and applications, such as bias created by machine learning and the related data science, or responsibility attribution problems that emerge from these methods and processes. Many of these issues do not only play out at an individual level, but also concern transformations in societies and economies. This is especially the case with AI-powered automation, which enables machines to take over tasks from humans.

This article gives a brief overview of some of the ethical issues and summarizes and discusses a number of challenges for near-future regulation in this area. The focus is on artificial intelligence applications that involve machine learning and data science.

2. Some ethical issues raised by artificial intelligence

Since AI and especially machine learning methods involve a process of data collection, processing, and sharing, a first issue – shared with many other digital technologies – concerns the question whether the privacy of individuals is respected and even whether they know that their data is collected at all. In the context of AI and data science these questions are especially urgent since often users do not know that AI is behind an application they use (e.g., an app on their phone) and since often data given in one context and one domain are then used by another party in another context and another domain, without the knowledge and consent of the people who gave their data.

Another well-known problem is data security: all these systems are networked and may be hacked for malicious purposes (e.g., cybercrime, cyberwar). The technology also relies on vulnerable material infrastructures: AI and other information systems are not entirely made of immaterial code but are embedded in material technological systems and require material infrastructures, which can be disrupted or destroyed.

Moreover, a problem that becomes especially relevant in the case of AI is attribution of responsibility. Since technologies cannot be responsible moral agents and are hence a-responsible, the only way to ensure responsible action is to make humans responsible. However, in technological action it is notoriously difficult to ascribe moral responsibility due to the so-called problem of "many hands": many people are involved in the often long causal histories that lead to a particular outcome. If there is a problem with the end result, say a recommendation, it is difficult to figure out who was responsible. And since AI is often part of a larger technological system and data histories, it is difficult to figure out if "the AI" caused the problem or some other part of the system. There are not only many hands but also many things.

Responsibility is especially problematic when people who use the systems are lured by the potential of the technology and use it without much hesitation, but are ignorant about most of the system and its history, for example how the data has been generated and combined. People using the systems are supposed to take responsibility but this becomes difficult if they don't know what they are doing.

* Prof. Dr. Mark Coeckelbergh is a full Professor of Philosophy of Media and Technology at the Department of Philosophy of the University of Vienna and the President of the Society for Philosophy and Technology (SPT).

Mark Coeckelbergh, Artificial Intelligence: Some ethical issues and regulatory challenges,
Technology and Regulation, 2019, 31–34 • https://doi.org/10.26116/techreg.2019.003 • ISSN: 2666-139X

But even experts don't always know everything, and this leads us to the problem of transparency and explainability. It is not always clear what is happening in the process, and this is especially the case for so called "black box" systems like machine learning that uses neural networks where technically an outcome (recommendation) cannot be traced back to a chain of decisions or reasoning as in decision tree models. Such systems are thus opaque. This is an ethical problem since people should have the right to know why a decision that affects them was taken. If a decision cannot be explained, this is unjust. Explainability is thus a moral requirement.

The problem of bias, furthermore, is especially challenging. Bias means that some individuals or groups are disadvantaged by the outcome of the system. Although problems of bias and discrimination have always been present in societies and cultures, the concern here is that the AI technology may perpetuate these and increase their impact. Bias is often unintended, but may be generated at various stages of the machine learning and data science process. Bias can arise in the selection of the data set, in the training dataset itself, in the algorithms used, in the application dataset, and indeed in wider society. Consider for example an AI that is trained on text data from the internet, which contains bias in the particular texts or even in the language (e.g., English). Perhaps bias cannot be avoided, in the sense that surely algorithms used for making decisions (e.g., about job applicants) are used for discriminating (e.g., between suitable candidates and others). But the question is always if a particular bias and discrimination is unjust and unfair. An answer to that question is not a merely technical question but an ethical and political one; it depends on our views of justice and on what kind of society we want.

Finally, in so far as AI is used for automation it also impacts work and the future of society. Many authors warn of unemployment and raise the question if a re-structuring of our social institutions is necessary (e.g., basic economy) to answer some of these challenges. This also makes us think about the political question *who* will decide about the technological future.

3. Addressing ethics of AI issues

While many policy makers that seek regulation of AI agree that something needs to be done in response to these ethical problems, they face a number of challenges. For a start, they need to answer the following questions: they need to figure out *what* should be done, justify *why* it should be done, *by whom* it should be done, and so on. For example, it is not easy to deal with the problem of bias: it is not clear what, exactly, should be done to avoid it as much as possible, and who should take action. And if existing regulation is seen as insufficient, new regulation should be justified: why is it needed, why is the existing regulation not enough? For example, in the case of data protection and privacy but also with regard to transparency and explainability, some argue that the European General Data Protection Regulation (GDPR)[1] instrument, which provides enforceable legislation, is sufficient; others argue that it does not provide enough protection against the risks of automated decision-making when it comes to explainability: there is only a right to information but this does not require full explainability.[2]

4. Guidance

So what about the future? The past year has seen a large number of policy documents that address ethics of AI, both from the public and the private sector. For example, already under the Obama presidency, the U.S. government published a report on the future of artificial intelligence[3] and last year many European countries published reports and strategies, for example the UK[4] and France[5]. Many documents propose trustworthy AI and explainable AI, and this has been reflected in supranational work on AI policy. In April 2018, the EU set up a new High-Level Expert Group on AI (HLEG AI) which has recently produced a document with ethical guidelines for AI (European Commission 2019). Earlier the European Group on Ethics in Science and New Technologies (EGE)[6] released a statement on AI which also proposes a number of principles. China and other major global players also have an AI strategy that includes ethics. For example, China has a development plan that recommends minimizing risk.[7] In addition, there also have been civil society actors commenting or campaigning with regard to AI, for example to ban autonomous weapons or to protect the privacy of citizens. And the Institute of Electrical and Electronics Engineers (IEEE), a large international technical professional organization, has taken a Global Initiative on Ethics of Autonomous and Intelligent Systems which has resulted in guidelines for ethical design.[8] And companies such as Google also published AI principles. They are not necessarily opposed to regulation; Apple's CEO Tim Cook has said that tech regulation is inevitable.[9] However, most industry players seem to prefer a minimal degree of regulation. This is a challenge for those who wish to move towards more substantial regulatory efforts.

Most policy proposals concerning AI ethics start from a number of ethical principles. For example, the HLEG AI starts from fundamental rights (human dignity, freedom of the individual, respect for democracy, justice and the rule of law, and citizens' rights) and a number of ethical principles, some of which are known from bioethics (the no harm principle, for example) but also explicability. These principles are relevant to AI in the form of machine learning: no harm requires that AI algorithms avoid discrimination, manipulation, and negative profiling, and explicability is interpreted as requiring that AI systems be auditable and comprehensible.[10]

However, this approach in terms of principles raises a number of challenges.

1 Regulation (EU) 2016/679 of the European Parliament and of the Council of 27 April 2016 on the protection of natural persons with regard to the processing of personal data and on the free movement of such data, and repealing Directive 95/46/EC (General Data Protection Regulation 2016 [OJ L 119, 4.5.2016, pp. 1–88].

2 Sandra Wachter, Brent Mittelstadt and Luciano Floridi, 'Why a Right to Explanation of Automated Decision-Making Does Not Exist in the General Data Protection Regulation' *International Data Privacy Law*, Volume 7, Issue 2, 1 May 2017, 76–99,

3 National Science and Technology Council Committee on Technology, 'Preparing For the Future of Artificial Intelligence' (Executive Office of the President, Office of Science and Technology Policy (OSTP) 2016).

4 House of Commons, 'Algorithms in Decision-Making, Fourth Report of Session 2017-19' (2018).

5 Cédric Villani, 'For a Meaningful Artificial Intelligence - Towards a French and European Strategy' (2018) https://www.aiforhumanity.fr/pdfs/Mission-Villani_Report_ENG-VF.pdf.

6 European Group on Ethics in Science and New Technologies (EGE), 'Statement on Artificial Intelligence, Robotics and "Autonomous" Systems' (European Commission, Directorate-General for Research and Innovation 2018).

7 'New Generation Artificial Intelligence Development Plan. (新一代人工智能发展规划) Translation Available at https://Flia.Org/Notice-State-Council-Issuing-New-Generation-Artificial-Intelligence-Development-Plan/ (State Council of China 2017).

8 https://ethicsinaction.ieee.org/

9 https://www.businessinsider.de/apple-ceo-tim-cook-on-privacy-the-free-market-is-not-working-regulations-2018-11

10 European Commission High-Level Expert Group on Artificial Intelligence (HLEG), 'Ethics Guidelines for Trustworthy AI' (European Commission 2019) https://ec.europa.eu/futurium/en/ai-alliance-consultation.

5. Challenges ahead

First, it is not clear if these expressions of concern for ethics of AI will actually lead to concrete regulation (when it comes to public actors) or concrete actions by corporations (private sector). While, for example, the European Commission set up procedures to stimulate uptake by stakeholders, there is no guarantee that this will actually happen. There is a risk that ethics are used as a fig leaf that helps to ensure acceptability of the technology and economic gain but has no significant consequences for the development and use of the technologies.

Second, even if stakeholders intend to do something with these documents, it is a challenge for regulation to move from more or less vague and abstract principles to more concrete methods, procedures, laws, and institutions. What are the concrete outcomes? Will there be new directives? New laws? Will there be a new agency that can monitor the implementation? While the European Commission document goes some way towards operationalization (further than any of the other documents I read), there is still a lot of work needed in this direction. It remains a huge challenge to bridge between abstract principles and concrete practices.

Third, one of these practices includes development and design of technologies; hence one may propose an ethics by design approach and similar measures. A proactive approach to technology ethics requires that ethics does not only come afterwards, by means of regulation after the technology is already developed, but that a regulatory framework is created to stimulate and (hopefully) ensure that ethics is already taken into account in earlier stages: in the development of the technology. For example, ethics by design could mean that it is required that traceability is ensured at all stages.[11] It is a challenge to think about how to technically implement ethics. For example, Winfield et al.[12] have called for implementing an 'ethical black box' in robots and autonomous systems which records data from sensors and the internal system; this could also be applied to AI. More generally, it is challenging to think about how to ensure explainability in technical ways. A related idea is responsible innovation[13], which requires that all kinds of stakeholders are involved in these earlier stages of development, potentially rendering the whole process more democratic and just.

These ideas also support the vision that regulation need not all be about banning things. We need a positive and constructive ethics of AI, which is not only about regulation in the sense of constraints but which also concerns the question of the good life and human and societal flourishing. Before thinking about concrete regulation, policy makers are challenged to develop a positive vision about where AI should take us.

6. Ethics by design and responsible Innovation

However, ideas such as ethics by design and responsible innovation and their implementation have their own barriers. It may be difficult to operationalize the general principles.

A. First, it is already great that explainability is operationalized as traceability in the HLEG guidelines, but what exactly does traceability mean? To find out what exactly should be done is itself a research question.
B. Second, ethics by design sounds great but it is not so easy to foresee the unintended consequences of new technologies at an early stage. More thinking needs to be done about concrete methodologies and techniques.
C. Third, it is hard to see how responsible innovation can really be implemented when there is a concentration of power in the hands of a relatively limited number of powerful actors, including a small number of large corporations: it seems that a handful of companies decide the future of AI.[14]
D. Fourth, can we really make fully explicit our values[15], given that ethical knowledge is partly tacit?
E. Finally, ethics by design, value sensitive design, responsible innovation, etc. work on the assumption that the technology will be developed[16]; is there also at least the possibility that the technology or the applications can be halted? How much room is there for deciding otherwise?

Fourth, there needs to be more interaction between legal and ethical expertise. For example, there are interesting questions with regard to which legal instruments can and should be used for dealing with problems of responsibility. For instance, whereas criminal law requires the intention to do harm, negligence asks the question whether a person was under a duty of care to prevent harm; this seems more applicable to AI and the people involved in its processes. Product liability, furthermore, does look at the fault of the person but has the company who produced the AI pay for damages, regardless of fault.[17] This could also be an interesting route to deal with responsibility issues. More generally, there needs to be a discussion about which legal instruments (existing or new) can and should deal with the ethical problems indicated.

Fifth, more generally, there is still a gap in understanding between people coming from the humanities and social sciences and those who have a technical background. A lack of interdisciplinarity can hinder the effectiveness of policy making in an area such as ethics of AI, when parties involved are constrained by their disciplinary understandings. Similarly, transdisciplinarity is needed in the sense that experts from academia need to reach out to (other) stakeholders and vice versa. We need to think about ways to bring together people and domains of knowledge and experience, not only in policy-making and professional life but also in the stage of education.

Sixth, it seems that given the nature of the technology, the problems are global and need to be addressed at a global level. But this is difficult when policy-making is largely happening at nation state level. How effective is it to take regulatory measures at the national level when the technology is developed and used across borders?

Finally, AI ethics policy is also a matter of priorities. There may be other technologies that also stand in need of regulation. And there may be national and global issues that also require our ethical and

11 Virginia Dignum and others, 'Ethics by Design: Necessity or Curse?', *Proceedings of the 2018 AAAI/ACM Conference on AI, Ethics, and Society - AIES '18* (ACM Press 2018) http://dl.acm.org/citation.cfm?doid=3278721.3278745 accessed 1 May 2019.

12 Alan FT Winfield and Marina Jirotka, 'The Case for an Ethical Black Box' in Yang Gao and others (eds), Towards Autonomous Robotic Systems (Springer International Publishing 2017)

13 René von Schomberg, 'Towards Responsible Research and Innovation in the Information and Communication Technologies and Security Technologies Fields' (European Commission 2011).

14 Paul Nemitz, 'Constitutional Democracy and Technology in the Age of Artificial Intelligence' DOI 10.1098/RSTA.2018.0089 - Royal Society Philosophical Transactions A

15 Paula Boddington, Towards a Code of Ethics for Artificial Intelligence (1st ed. 2017 edition, Springer 2017).

16 Kate Crawford and Ryan Calo, 'There Is a Blind Spot in AI Research' (2016) 538 Nature 311.

17 Jacob Turner, Robot Rules - Regulating Artificial Intelligence (Palgrave Macmillan 2019).

political attention, such as social-economic injustices and climate change. A good AI policy that aims to be ethical needs to address this question of priorities, which is an ethical and political question.

If these barriers can be overcome, there is a chance for effective and good regulation of AI in an ethical direction and, more generally, an AI future that we want.

04

Technology and Regulation

Legality and Democratic Deliberation in Black Box Policing

Markus Naarttijärvi*

Policing, legality, Rule of Law, technology, black box policing, democracy, surveillance, machine learning

markus.naarttijarvi@umu.se

The injection of emerging technologies into policing implies that policing mandates in law may become mediated and applied through opaque machine learning algorithms, artificial intelligence, or surveillance tools – contributing to a form of 'black box policing' challenging foreseeability and clarity and expanding discretionary legal spaces. In this paper, this issue is explored from a constitutional and rule of law perspective, using the requirements of qualitative legality elaborated by the European Court of Human Rights and the implicit democratic values that they serve. Placing this concept of legality into a wider theoretical framework allows legality to be translated into a context of emerging technology to maintain the connections between rule of law, democracy, and individual autonomy.

1. Introduction

1.1 Governing by, and through, technology

Governing is increasingly mediated through digital technology. This is visible in everyday citizen-government interactions, such as online applications for government benefits, income tax declarations and other common e-government services. The digitally mediated nature of governing becomes even more apparent in the face of algorithmic decision-making, where big data and machine learning form a basis for the application of government power and authority.[1] Moreover, the classification of individuals through the observation of their digital footprints is increasingly establishing itself as a governmental shorthand of power, potentially forming the basis for both coercive actions and lethal force.[2] While often discussed in terms of the potential for interferences with privacy or data protection rights, these developments also challenge more fundamental legal values; given the importance of digital technologies in the current exercise of government power, the technologies themselves become crucial for the analysis of whether or not legality as a basic rule of law value is upheld. Legality, understood here in a constitutional context, implies that the exercise of government power should have a basis in law. In a modern understanding – influenced by rule of law values and human rights adjudication – legality also establishes that this legal basis must reach a certain quality; to ensure the accessibility and clarity of law, enable foreseeability, and limit government discretion.[3] As will be shown, these qualitative aspects of legality, as elaborated most clearly by the European Court of Human Rights, also fulfil other, more implicit but equally important, democratic values.

The hypothesis of this paper is that technology adds obscurity to the exercise of law and government power. This obscurity may in many contexts affect the ability to uphold legality as a rule of law value and as a normative limit to government power. While uncertainty is not uncommon in law, it is traditionally perceived as an issue connected to the clarity of legal rules as such and the often-unavoidable indeterminacies of human language that law is expressed through, or such generalisations that are intentionally included to ensure a certain flexibility.[4] Technology, however, adds a different layer of obscurity as the effect of law and the exercise of government power is mediated through a layer of coded norms, logic and presumptions that are external to law and that may be unforeseeable to both legislators and citizens. Technology, as will be shown, may simultaneously act as a driver of vague or indeterminate legislation and inject indeterminacy into an otherwise clear and foreseeable language of law. This raises issues not only with legality, but also with societal values that legality serves, such as the separation of powers, individual autonomy, and democratic legitimacy.

* Markus Naarttijärvi is an associate professor of law at the department of law at Umeå university, Sweden.

Received 25 June 2019, Accepted 22 Oct 2019, Published: 1 Nov 2019

1 See Andrew D Selbst, 'Disparate Impact in Big Data Policing' (2018) 52 *Georgia Law Review* 109; Mireille Hildebrandt, 'Proactive Forensic Profiling: Proactive Criminalization?' in R Anthony Duff and others (eds), *The boundaries of the criminal law* (Oxford University Press, 2010); Virginia Eubanks, *Automating Inequality: How High-Tech Tools Profile, Police, and Punish the Poor* (St Martin's Press, 2018); Andrew Guthrie Ferguson, *The Rise of Big Data Policing: Surveillance, Race, and the Future of Law Enforcement* (NYU Press, 2017).

2 See Kevin D Haggerty and Richard V Ericson, 'The Surveillant Assemblage' (2000) 51 *The British Journal of Sociology* 605; Michel Foucault, *Discipline and Punish: The Birth of the Prison* (Vintage Books, 1995); Paul De Hert and Serge Gutwirth, 'Data Protection and Law Enforcement: Opacity of the Individual and Transparency of Power' in Anthony Duff, Serge Gutwirth and Erik Claes (eds), *Privacy and the Criminal Law* (Intersentia, 2006).

3 See further section 3 below.

4 Cf. Timothy AO Endicott, *Vagueness in Law* (Oxford University, 2000) 160–164.

The implications of technology are of particular concern in relation to policing, as a context of government power that is subject to detailed regulation given its implications for individual rights, while it is simultaneously an activity which is characterised by significant amounts of autonomy and discretion for both officers and police authorities.[5] As such, technology can be applied in policing through these discretionary spaces while having significant effects on the exercise of power in practice – creating in effect a form of *black box policing* affecting the ability of both citizens and legislators to understand the scope and impetus of police actions and the role technology has played in shaping them. Policing is also an area subject to intense public and political pressure to 'get the job done', which further incentivises the use of technology to reach efficiency targets.[6] Consequently, policing is an area of law where the implications of technology in terms of mediating law and policy into practical effects for individuals may carry tangible and far-reaching implications. The examples provided in the policing context may therefore illustrate implications of obscurity due to technologically mediated governing for both legality and democracy which are relevant for other contexts as well. It may also lay the foundation for an analysis of how legality as a component of rule of law may be translated into a context of technologically mediated governing to preserve such values that underpin legality.

First, however, something should be said about the term *technologically mediated governing*. I use this term here as a shorthand for a behind-the-scenes normative layer of code and data that change the implications of governing through law and government decisions. There are somewhat similar concepts used by other authors carrying other implications. In his analysis of the role of technology as a tool of governing Brownsword uses the term 'technological management', referring to how technology is used normatively to restrict or reduce existing human possibilities by making rule breaking technologically impossible; a simple example is technologically ensuring that cars stop at red lights rather than relying on norms to encourage or coerce drivers to do so.[7] I use the term technologically mediated governing here to instead signify how the application of a certain technology alters (i.e. mediates) the implications of governing through law, rather than through technology as such. This may in some instances include technological management as conceptualised by Brownsword, however, technologically mediated governing is not dependent on the restriction or reduction of human possibilities through technology itself.[8] In other words, the interest is not so much how technology serves to ensure individual compliance, as how the exercise of government power mediated by technology affects legality. In this sense, I approach the technologically mediated nature of governing from a perspective that is similar to the concept of digitisation as defined by Yoo *et al.* as 'the transformation of existing socio-technical structures that were previously mediated by non-digital artefacts or relationships into ones that are mediated by digitized artefacts and relationships with newly embedded digital capabilities'.[9] This goes beyond the mere technical process of digitisation of analog information and 'involves organizing socio-technical structures with digitized artefacts [and] the reconfiguration of broader socio-technical structures that were previously mediated by non-digital artefacts'.[10] The use of these technologies also implies, as noted by Latour, the mobilisation of 'moves made elsewhere, earlier, by other actants'.[11] This entails that technologies used in policing will effectuate the values, choices, and norms embedded in those technologies at an earlier date. In other words, the mediation of technology will not only alter implications of law through its interpretation into new contexts, or the new possibilities afforded by the technology,[12] but also through a form of normative refraction which occurs as the legal norms interact with the embedded values, choices, and norms of the technology used.

In the rest of this first section, I will underpin the importance of technology as a tool of governing through conclusions drawn in existing research. In section 2, I will point to examples from the policing context where technologically mediated governing challenges legality. These examples will serve as a background for a broader analysis of legality in section 3, first as a more abstract value, then as a normative requirement as applied in the case law of the European Court of Human Rights (ECtHR). In section 4, I will argue for a broader reading of legality which implicitly serves democratic values. Finally, in section 5, I will sketch out an understanding of legality that highlights the importance of upholding both legal and democratic values in the face of emerging technology and provide some tentative recommendations on how to approach its application.

1.2 The importance of features, code, and data

The importance of technology for governing has become apparent since the rise of the network society.[13] As Lawrence Lessig has noted, we embed different values when constructing code and choosing different technological architectures, and the decisions made regarding these same codes and architectures enable control from whatever sovereign that does the coding.[14]

> [I]f in the middle of the nineteenth century the threat to liberty was norms, and at the start of the twentieth it was state power, and during much of the middle twentieth it was the market, then my argument is that we must come to understand how in the twenty-first century it is a different regulator – code – that should be our current concern.[15]

5 Elizabeth E. Joh uses the term 'surveillance discretion' to refer to the far-reaching discretion of the police in deciding who to investigate and focus their attention on. See Elizabeth E Joh, 'The New Surveillance Discretion: Automated Suspicion, Big Data, and Policing' (2016) 10 *Harvard Law and Policy Review* 15, 16. See also Selbst (n 1) 119, who comment that '[p]olice act with incredible discretion. They choose where to focus their attention, who to arrest, and when to use force. They make many choices every day regarding who is a suspect and who appears to be a criminal.'

6 Cf. Lena Landström, Niklas Eklund and Markus Naarttijärvi, 'Legal Limits to Prioritisation in Policing – Challenging the Impact of Centralisation' (2019) *Policing and Society* (online pre-print).

7 Roger Brownsword, 'In the Year 2061: From Law to Technological Management' (2015) 7 *Law, Innovation and Technology* 1, 8.

8 As such, the interest here — to use the same example of the red light — is rather how technology may be used to either identify persons who did not stop at the red light and then use law to sanction them, or more proactively to identify who is more likely not to stop at the red light and then use existing government powers to control or coerce them in their car use.

9 Youngjin Yoo and others, 'Unbounded Innovation with Digitalization: A Case of Digital Camera', 2010 *Annual Meeting of the Academy of Management* (2010) 4.

10 Yoo and others (n 9) 4. As Yoo et al looked at digitisation of products, these artefacts would in this context instead be the digitisation of government powers and methods.

11 Bruno Latour, 'On Technological Mediation' (1994) 3 *Common Knowledge* 29, 52. See also Don Ihde, *Technology and the Lifeworld* (Indiana University Press, 1990) 49, stating that 'for every revealing transformation there is a simultaneous concealing transformation of the world, which is given through a technological mediation. Technologies transform experience, however subtly, and that is one root of their non-neutrality'.

12 Cf. Peter-Paul Verbeek, *Moralizing Technology – Understanding and Designing the Morality of Things* (University of Chicago Press, 2011) 5.

13 For the use of this term, see Jan van Dijk, *The Network Society* (Sage Publications, 2012).

14 Lawrence Lessig, *Code: Version 2.0* (Basic Books, 2006) 77, 114.

15 Lessig (n 14) 121.

Given the importance of architecture, it is, Lessig holds, important not to ignore this type of regulatory modality or accept it as given – rather, it needs to be taken into account in the making of law, and the technological responses to law must be predicted.[16]

In a similar vein, Hildebrandt aptly uses the term *affordances*, borrowed from biology, to explain how 'technologies afford certain behaviours that would otherwise have been impossible, or do not afford certain behaviours that were available before the technology was in place'.[17] From this point of departure, she argues that criminal justice has been afforded a more actuarial approach where the focus is placed on profiling and the characteristics and calculated risk a person represents, rather than the actual actions of that person as such.[18] Such actuarial justice may also be represented by the increased emphasis on intelligence-led policing (ILP), focusing on patterns and predictability rather than approaching crime on a case-by-case basis.[19]

As such, technology will affect *what* law governs, but also *how* law governs. For example, before the advent of digital networking, the idea of massive interception and automated processing of telecommunications was scarcely afforded by the available technology.[20] Given the increased availability of data and the development of processing power and software to automatically process these data, the concept of massive, or bulk, interception has increasingly become afforded by technology, and as such a clear focus of government surveillance efforts and legislation in the last decades. While modern conceptions of terrorism following 9/11 have acted as drivers of this type of surveillance, the interaction between developments in the security paradigm and the technological developments of data processing has acted as a catalyst to enable the rise of the modern surveillance state.[21]

The affordance of new methods of governing within the field of policing brings us to one of the main issues in relation to legality, namely the potential for technological obscurity – i.e. the way the injection of technology can cloud the implications and effects of legal mandates and policing methods, with potential effects for both the accessibility and foreseeability of law.

2. Delineating the black box of policing

As previously mentioned, the hypothesis of this article is that technology adds obscurity to the application of law and government power. In this section, I will establish the further basis for this hypothesis and outline four ways in which technology either expands discretionary spaces in ways that are opaque for persons outside of a police force, or injects obscurity into existing methods of policing, thereby shifting the practical and regulatory environment where police authorities act. This is not an exhaustive list of possible concerns, but represents such areas of concern that have been either highlighted in previous research or that present a *prima facie* challenge to the ideals of qualitative legality, as I will soon describe further.

2.1 Avoiding regulatory negotiation: discrete and direct application of technology

In many contexts, executive agencies and other government organs, including law-enforcement authorities, are dependent on law-makers to arbiter and decide where the interests of public authorities collide with those of private interests or individuals. This is the case when the law requires private entities to assist the police in inquiries or provide material support such as enabling and assisting the police in the surveillance of phone networks. A clear example of this is how the EU data retention directive – while in force – created a responsibility for EU member states to enact legislation which required private telecommunications providers to retain communications data for law enforcement purposes.[22] When enacting these rules, law-makers – and by extension courts – are forced to balance public and private interests, while keeping in mind such constitutional rules and limits that may provide a proverbial thumb on the scale in certain contexts.[23]

The situation is however different when authorities can achieve their aims by more direct and discrete means. Technologies such as IMSI-catchers (a piece of equipment masquerading as a mobile base station, capturing information about nearby mobile equipment) upsets this balance by allowing – in the practical sense – authorities (as well as private parties) to monitor communications and surrounding devices without going through telecommunications providers.[24] The implication of direct and discrete applications of technology is that the very practical need for the legislature to enable the application of a certain technology within government agencies is reduced or eliminated. There are no communication providers to convince or coerce into cooperation when using an IMSI-catcher, as the technology affords direct surveillance to whomever has access to the equipment in question. As such, the nature of the technology in conjunction with efficiency demands invites authorities to apply the technology, even when the regulatory environment may not support it.[25] The reduced need for legislators to practically enable surveillance through legal norms thus affects the impetus for basing such surveillance on clear and foreseeable legal rules. This is exacerbated by the fact that such technologies are more difficult to challenge in court,

16 Lessig (n 14) 126, 129.

17 Hildebrandt (n 1) 121.

18 Hildebrandt (n 1) 124–277.

19 See Nick Fyfe, Helene Oppen Gundhus and Kira Vrist Rønn, *Moral Issues in Intelligence-Led Policing* (2017) 1–20; Nick Tilley, 'Modern Approaches to Policing: Community, Problem-Oriented and Intelligence-Led' in Tim Newburn (ed), *Handbook of Policing* (2nd edn, Willan Publishing, 2008).

20 Though a more manual and resource intensive form of massive surveillance was implemented in many countries during the second world war, in Sweden for example, it has been estimated by the Swedish security service that over 11 million telephone calls were subject to interception during the war years, see Säkerhetspolisen, *Säkerhetspolisens Årsbok 2013* (Swedish Security Service, 2013).

21 See generally Markus Naarttijärvi, *För Din Och Andras Säkerhet – Konstitutionella Proportionalitetskrav Och Säkerhetspolisens Preventiva Tvångsmedel* (Iustus förlag, 2013).

22 Directive 2006/24/EC of the European Parliament and of the Council of 15 March 2006 on the retention of data generated or processed in connection with the provision of publicly available electronic communications services or of public communications networks and amending Directive 2002/58/EC.

23 As it did in the cases from the CJEU invalidating the data retention directive, see Joined Cases C 293/12 and C 594/12, Digital Rights Ireland Ltd v Minister for Communications, Marine and Natural Resources and Others and Kärntner Landesregierung and Others [2014], Judgment of the Court (Grand Chamber), 8 April 2014 (ECLI:EU:C:2014:238).

24 See Stephanie K Pell and Christopher Soghoian, 'Your Secret StingRay's No Secret Anymore: The Vanishing Government Monopoly over Cell Phone Surveillance and Its Impact on National Security and Consumer Privacy' (2014) 28 *Harvard Journal of Law & Technology* 1, 9. They described this as "direct and unmediated" surveillance technologies, I use the term direct and discrete here to avoid confusion with the term technologically mediated governing.

25 The Swedish police authority has, incidentally, been using IMSI-catchers since – at least – 2005, without a mandate in law, in violation of EU-law and the European Convention on Human Rights. As the method is secret, the difficulty is however to establish legal standing to challenge this in courts. See Markus Naarttijärvi, 'Swedish Police Implementation of IMSI-Catchers in a European Law Perspective' (2016) 32 *Computer Law & Security Review* 852.

as the details of their implementation and use are known primarily within the agencies using them.[26]

While there has been increased focus on the need for judicial warrants to legally use IMSI-catchers in criminal investigations, the direct and discrete nature of this and similar technologies – like *Finfisher*-like hacking tools – may create a significant delay in the application of judicial controls. Law-enforcement agencies can in effect take advantage of the legal uncertainty surrounding the method, the covert nature of its use, and the direct and discreet nature of the measure to preclude legal challenges. In Canada, law enforcement agencies have been reported to accept plea-deals rather than risk that the use of IMSI-catchers in the investigations become known through discovery and subject to legal challenges.[27] In the United States, secrecy surrounding the same technology has also been attributed to non-disclosure agreements,[28] which brings me to my next point.

2.2 Outsourcing policy choices and acceded secrecy: proprietary private sector product development

A second way in which technology creates obscurity is through what could be described as an 'outsourcing of choice' to the private sector and the associated proprietarisation and confidentiality of policing technology and methods.

In her important work within this area, Elizabeth E. Joh has analysed the 'undue influence of surveillance technology companies on policing' in the United States. She points to how private companies within the surveillance industry make choices in their product development that will influence important aspects of how technology is applied – implicitly affecting policy choices by determining the available choices.[29] This is interconnected with the previously mentioned discrete and direct nature of many technologies, allowing them to function and develop without the law properly mediating between private and public interests. The feature set of such products and their future development may be primarily adapted to larger jurisdictional markets, making the implementation of such products into police forces in jurisdictions for which the product has not been adapted problematic as the product may not fit its particular legal framework. From the point of view of obscurity, a further concern with this proprietarisation is that the features and capabilities of these commercial products may be covered by confidentiality agreements between the producer and the purchasing law enforcement agency, or subject to claims of trade secret protection which may limit the effect of information requests.[30] This may severely interfere with transparency, giving little or no public insight into the actual effects and implications of police powers. It may also hinder legal challenges to their implementation in a certain jurisdiction as police authorities may have agreed to limit the exposure of the technology in court proceedings.

2.3 Changing the equation: the use of existing data in novel ways

A subtler way that technology can mediate the exercise of government power is through emerging ways of analysing and operationalising data already available to law enforcement. Either through new applications of these data, or their use on a scale that was not factored into the original legislative calculation. This is clearly accentuated by the rise of data mining, big data policing, and actuarial justice.[31]

The point here is that technologies like data mining can shift the basic paradigm of policing. Whereas traditional policing is largely incident driven – responding to incoming reports, emergency calls and events,[32] data mining affords law enforcement agencies to adopt more forward-looking approaches where they act on their own initiatives to try to prevent or mitigate future undesirable acts. As Selbst has pointed out, '[d]ata mining allows police to operate unconstrained by theory, finding correlations without worrying why they work'.[33] Causality, in this context, is not as interesting as these correlations, as Joh has noted:

> In criminal investigations, it may not be necessary to know why certain patterns of driving, purchasing, or movement are associated with crime if the police can claim a high correlation between the two. A high degree of correlation itself might provide justification for heightened police attention.[34]

This may also shift the basis for when police powers are used and may circumvent the logic underpinning due process rights surrounding the use of such powers. Traditional concepts such as *reasonable cause* or *reasonable suspicion* regarding individual suspects of a crime that has been committed are difficult to apply in relation to persons finding themselves in an area designated as a potential future crime hotspot, or who have been placed on a 'heat-list' as likely to commit future crimes.[35] Here, the issue is both connected to the translation of risk to traditional evidentiary requirements surrounding coercive powers – i.e. whether a statistical risk is enough to warrant coercive measures – and if the data themselves are trustworthy or carry a potential for hidden biases. However, to a large extent the issue of big data policing relates to the potential for inequitable distribution of police *attention* based on such data.[36] The focusing on police attention is traditionally an issue of police discretion that is largely unregulated, but which carries with it inherent issues of equality and fairness that may be exacerbated by the application of emerging technologies of algorithmic decision-making.[37] Several researchers have pointed to the potential for predictive policing to create feedback loops whereby existing inequalities in the distribution of police attention create a biased data set which focuses even more attention to certain areas or individuals in the future – attention that may not be warranted

26 See section 2.2 below.
27 See in the Canadian context Colin Freeze, 'Guilty Pleas End Risk of Revealing RCMP Surveillance Technology' *The Globe and Mail* (30 March 2016) https://www.theglobeandmail.com/news/national/guilty-pleas-scuttle-hearing-that-risked-revealing-rcmp-surveillance-technology/article29430116/ accessed 11 October 2019.
28 Brad Heath, '200 Imprisoned Based on Illegal Cellphone Tracking, Review Finds' *USA Today* (14 December 2016) https://eu.usatoday.com/story/news/2016/03/31/200-imprisoned-based-illegal-cellphone-tracking-review-finds/82489300/ accessed 11 October 2019.
29 See Joh (n 5) 113–114, discussing police body cameras.
30 See Joh (n 5) 126–126; Selbst (n 1) 189.
31 See Ferguson (n 1); Eubanks (n 1).
32 See Landström et al (n 6).
33 Selbst (n 1) 129.
34 Joh (n 5) 21.
35 Selbst (n 1) 137; Joh (n 5). See also for a more concrete example Vicky Sentas and Camilla Pandolfini, *Policing Young People in NSW: A Study of the Suspect Targeting Management Pan* (Youth Justice Coalition NSW, 2017).
36 Vlad Niculescu-Dincă, 'Towards a Sedimentology of Information Infrastructures: A Geological Approach for Understanding the City' (2018) 31 *Philosophy & Technology* 455, 468.
37 Joh (n 5) 18–19. Niculescu-Dincă (n 36). The inequality of attention may also result in certain victims, not represented in available data, becoming marginalised, see Jonas Lerman, 'Big Data and Its Exclusions' (2013) *Stanford Law Review* [online] https://www.stanfordlawreview.org/online/privacy-and-big-data-big-data-and-its-exclusions/ accessed 1 November 2019.

for the end result of actually preventing crime.[38] Application of data analysis tools may also create what Niculescu-Dincă has described as a sedimentation of design-choices – 'design choices are covered by sediment and thereby invisible, and the prejudices become rock solid in the working routines of the local police. In this way, they can induce a perception of objectivity towards the enacted community, affecting their presumption of innocence'.[39] Another point that has been highlighted by Lyria Bennett Moses and Janet Chan is how algorithmic prediction in policing rests on several assumptions that should be open to challenge, such as the data accurately reflecting reality, that the future will be like the past, and that algorithms are neutral.[40] Some of these assumptions will also negatively affect the transparency and accountability of the process because they are inherent and poorly understood.[41] Given that these assumptions are built into the idea of predictive policing as such, they are sedimented as well, hidden behind software features, and affecting the technological mediation of policing.

The main point here from the point of view of legality is that the application of these new analytic technologies in policing is not necessarily tied to express competences in law, but rather to the changing implication of existing discretionary spaces or areas of legislative inactivity. For example, social media posts are public and consequently law enforcement access to collect and analyse such posts may on the one hand be comparable to observing their surroundings – indeed one might ask: why the police should have less possibilities of observing online discourses than an everyday citizen? On the other hand, law enforcement access to social media posts entails issues that challenge that analogy. Unlike general observations of what happens in the physical world, the police can collect, aggregate, and analyse vast quantities of social media postings in a way which the observation of the physical world does not (yet) allow. As such, social media posts can provide data for analyses of social networks of citizens and afford semantic and mathematical analysis on a vast scale that creates real world implications for the exercise of police powers.[42] Existing commercial software can, for instance, allow police to assign 'threat scores' to persons or addresses in advance of responding to emergency calls, or attempt to identify active gang members. This in turn may change the way police behave and respond to calls, which the police claim can lead to a safer responses to incoming calls, whereas opponents claim the opaque and rough calculus of the software may lead to mistakes which implicitly increases the risk of citizens facing unnecessary force.[43]

38 See Annette Vestby and Jonas Vestby, 'Machine Learning and the Police: Asking the Right Questions' [2019] *Policing: A Journal of Policy and Practice paz035*; Selbst (n 1) 13, 27; Danielle Ensign and others, '*Runaway Feedback Loops in Predictive Policing*' [2017] arXiv:1706.09847 [cs, stat] http://arxiv.org/abs/1706.09847 accessed 14 August 2019; Bernard E Harcourt, *Against Prediction: Profiling, Policing, and Punishing in an Actuarial Age* (University of Chicago Press, 2007) 147–160.

39 Niculescu-Dincă (n 36) 465.

40 Lyria Bennett Moses and Janet Chan, 'Algorithmic Prediction in Policing: Assumptions, Evaluation, and Accountability' (2018) 28 *Policing and Society* 806, 809–815.

41 Bennett Moses and Chan (n 40) 818.

42 Joh (n 5) 24–26.

43 Selbst (n 1) 137; Joh (n 5) 24–26. See also Brent Skorup, 'Cops Scan Social Media to Help Assess Your "Threat Rating"' (Reuters Blogs, 12 December 2014) http://blogs.reuters.com/great-debate/2014/12/12/police-data-mining-looks-through-social-media-assigns-you-a-threat-level/ accessed 14 August 2019; Justin Jouvenal, 'The New Way Police Are Surveilling You: Calculating Your Threat "Score"' *Washington Post* (10 January 2016) https://www.washingtonpost.com/local/public-safety/the-new-way-police-are-surveilling-you-calculating-your-threat-score/2016/01/10/e42bccac-8e15-11e5-baf4-bdf37355da0c_story.html accessed 14 August 2019.

In sum, given that the application of these new technologies may take place without new legislative action, the risks and benefits that they bring may never have been the subject of any democratic deliberation. Yet the practical effects they yield may be substantial – forming the basis for a potentiality of both structural and individualised discrimination, coercive measures, and the translation of risk assessments into practical effects.

2.4 Open code versus algorithmic black boxes

Beyond the impact of emerging technologies on the discretionary spaces of policing powers, their adoption may also obscure the regulation of policing powers as such. If we accept Lessig's idea that code and architecture regulate, he further argues that this type of regulation will affect transparency. It allows the state to hide a regulatory agenda by pursuing it though indirect regulation.[44] As such, it may serve to render regulation – and by extension – the extent of government powers, invisible. Thus, the code that regulates becomes extremely important, and the transparency of that code may be crucial to the maintenance of overall foreseeability and transparency of power. Consequently, Lessig's solution in this regard was the use of open code.[45] Allowing access to the code would help with transparency and open up this type of regulation to scrutiny.

However, since Lessig articulated these arguments, the increased emphasis on 'big data', machine learning algorithms, and AI has highlighted the difficulty of achieving transparency through accessible code. For instance, the logic behind deep learning neural networks is not necessarily comprehensible even for the coders creating them – nor the officers applying them –, awarding them their nickname of 'black boxes'.[46] Indeed, the point of deploying such neural networks is to achieve a better prediction rate in their application than a human could accomplish and regardless of human comprehension of the logic.[47] In this process, the ability of AI or machine learning systems to generate unexpected or 'emergent' results may be regarded by designers as a significant competitive advantage.[48] Certain authors have challenged this idea of obscurity – pointing to the way the design process as a whole can provide some clarity,[49] but the general concern of lacking transparency persists. Furthermore, machine learning algorithms alter their own algorithmic logic in response to new data, continuously developing their prediction model after each new data point.[50] As a result, the underlying code is always evolving, and any transparency of the code will be momentary and fleeting. Finally, and importantly, the ability to predict the effect of a specific algorithm by looking at the code is limited as it will depend on the data it is fed and the quality of those data. Contrary to popular belief, like code – data are rarely neutral, instead they tend to reflect the inherent biases present in whatever environment they originate from. Non-discriminatory code may still produce discriminatory results if the data it is

44 Lessig (n 14) 135-136.

45 Lessig (n 14) 128, 139.

46 See generally Frank Pasquale, *The Black Box Society: The Secret Algorithms That Control Money and Information* (Harvard University Press, 2015).

47 Brent Daniel Mittelstadt and others, 'The Ethics of Algorithms: Mapping the Debate' (2016) 3 *Big Data & Society* 1, 6; Jenna Burrell, 'How the Machine "Thinks": Understanding Opacity in Machine Learning Algorithms' (2016) 3 *Big Data & Society* 1, 10.

48 Matthew U Scherer, 'Regulating Artificial Intelligence Systems: Risks, Challenges, Competencies, and Strategies' (2016) 29 *Harvard Journal of Law & Technology* 353, 365.

49 Joshua A Kroll, 'The Fallacy of Inscrutability' (2018) 376 *Philosophical Transactions of the Royal Society A: Mathematical, Physical and Engineering Sciences*.

50 Jenna Burrell, 'How the Machine "Thinks": Understanding Opacity in Machine Learning Algorithms' (2016) 3 *Big Data & Society* 1, 5.

fed contains a discriminatory bias, either as a result of a biased data source or a non-representative data set.[51]

The response from industry and academia have primarily been centred on countering these problems through the development of AI ethics. While ethical standards relating to AI are important in their own right, ethics is not a panacea. Indeed, the tendency to focus on ethics may risk delaying the activation of democratic structures and the regulation through law, instead relying on soft norms and code to govern the permissible extent of the functions and applications of AI.[52] Going back to the conceptualisation of Kantian ethics, any ethical action must first be legal, indicating a priority of considerations that indicate that ethics should be a complementary, rather than a first order concern in the management of the issues relating to AI.[53] Simultaneously, utilitarian ethics are prevented in many contexts by higher-order legal norms which explicitly express Kantian norms and do not allow for cost-benefit analysis with respect to individual rights.[54]

Consequently, to the extent that an otherwise clear and accessible law facilitates the adoption and use of non-transparent code in decision-making or the exercise of public power, the operation of law will be determined by inaccessible and potentially non-explainable factors – implicitly and indirectly challenging the ability of upholding legality. This brings us to what this concept of legality implies and the extent to which it can tackle the issues highlighted so far.

3. Conceptualising legality

3.1 A theoretical basis for qualitative legality

Legality as a rule of law value is a cornerstone of the modern democratic constitutional order. As an ideal of ruling through and under the law, legality has a long, albeit not straightforward, history in Europe dating back to Ancient Greece and Rome.[55] Today it is well established, at least in a European constitutional context, that the principle of legality may extend beyond a mere requirement that an exercise of government power has a formal basis in law. As will be shown, this wider understanding of legality, which I will refer to as qualitative legality, is influenced by principles of constitutionalism as well as legal theory. It adds several qualitative requirements to the law in question, for instance accessibility, clarity, precision, non-retroactivity, and a general application.[56] The impact of these requirements can be seen most clearly in relation to legal rules which limit fundamental rights where the formal concept of legality is supplemented with a more substantive understanding that focuses on the rule of law qualities of the legal rules. This development has become clearly apparent in the case law of the European Court of Human Rights (ECtHR, or 'the court') in its interpretation of the European Convention on Human Rights (ECHR).[57] Consequently, the same principles are also implicit in the EU Charter of Fundamental Rights.[58] As important rule of law values, they can also be found in the definition of the rule of law articulated by the Council of Europe's Venice Commission.[59]

While subject to development in recent years, qualitative legality is not a new idea as such, nor is it conceptually limited to the context of limitations of rights or within the area of criminal law where matters of legal certainty are most acute. The values implicit in qualitative legality have indeed been expressed more generally within jurisprudence as aspects of an *inner morality of law*, given how they act as internal legal modes of rationality in the absence of which we may question whether a legal system can be seen to exist at all. This understanding has been underpinned by the direct relationship between these qualitative requirements and the ability of law to govern the behaviour of individuals; in the absence of foreseeability, individuals cannot understand what the law requires of them and thus are not able to conform to these requirements.[60]

Most discussions on qualitative legality (or similar concepts by different names) have in common this underpinning of legal authority and legitimacy through the individual's ability to ascertain and understand what is expected of her. As such, legality derives its value largely from the point of view of the individual, where it forms a bastion against unrestricted or arbitrary government power and acting as a precondition for individual freedom and autonomy.[61] In this sense, the qualitative requirements have also been described as something akin to a contractual transaction; if the individual is expected to follow the wishes of the legislator, it is no more than right that the individual can also ascertain what those wishes are and rely on a reasonable interpretation of their legal expression.[62]

51 See Solon Barocas and Andrew Selbst, 'Big Data's Disparate Impact' (2016) 104 *California Law Review* 671; Danielle Keats Citron and Frank Pasquale, 'The Scored Society: Due Process for Automated Predictions' (2014) 89 *Washington Law Review* 1; Safiya Umoja Noble, *Algorithms of Oppression: How Search Engines Reinforce Racism* (New York University Press, 2018); Bennett Moses and Chan (n 40); Selbst (n 1).

52 See Ben Wagner, 'Ethics as an Escape from Regulation. From "Ethics-Washing" to Ethics-Shopping?' in Emre Bayamlıoglu and others (eds), *Being Profiled: Cogitas Ergo Sum – 10 Years of Profiling the European Citizen* (Amsterdam University Press, 2018); Paul Nemitz, 'Constitutional Democracy and Technology in the Age of Artificial Intelligence' (2018) 376 Philosophical Transactions of the Royal Society A: Mathematical, Physical and Engineering Sciences.

53 Immanuel Kant, *Grundlegung Zur Metaphysik Der Sitten* (Hoefenberg 2016).

54 For example, the right to human dignity as expressed in art. 1 of the EU Charter of Fundamental Rights, see Catherine Dupré, 'Art 1 – Human Dignity' in Steve Peers (ed), *The EU charter of fundamental rights: a commentary* (Hart [u.a], 2014).

55 Brian Z Tamanaha, *On the Rule of Law: History, Politics, Theory* (Cambridge University Press, 2004) 7-14.

56 See Andrei Marmor, *Law in the Age of Pluralism* (Oxford University Press, 2007) 6–7; Lon L Fuller, *The Morality of Law* (Yale University Press, 1964) 33–95.

57 See Geranne Lautenbach, *The Concept of the Rule of Law and the European Court of Human Rights* (Oxford University Press, 2013); David J Harris, M O'Boyle and Colin Warbrick, *Harris, O'Boyle & Warbrick: Law of the European Convention on Human Rights* (Oxford University Press, 2014) 506-509; Mattias Derlén, Johan Lindholm and Markus Naarttijärvi, *Konstitutionell Rätt* (Wolters Kluwer Sverige, 2016) 281-284, see further section 4 below.

58 Cf. Sacha Prechal and Steve Peers, 'Article 52 – Scope of the Protected Rights' in Steve Peers and others (eds), *The EU Charter of fundamental rights: a commentary* (Hart Pub Ltd, 2014) 1473. Though the ECJ has yet to put its foot down despite multiple references by advocate generals, the qualitative requirements should at the very least apply in relation to rights corresponding to the ECHR. See also Robert Schütze, *European Constitutional Law* (Cambridge University Press, 2016) 447, suggesting the ECJ applies a material rather than formal concept of law.

59 Venice Commission, *Report on the Rule of Law* (Venice Commission, 2011) 003rev-e, 41 & 44.

60 See Fuller (n 56) 33–95; Marmor (n 56) 6–7.

61 See Tamanaha (n 55) 34-35; Friedrich A von Hayek, *The Constitution of Liberty: The Definitive Edition* (University of Chicago Press, 2011) 320; TRS Allan, 'The Rule of Law' in David Dyzenhaus and Malcolm Thorburn (eds), *Philosophical Foundations of Constitutional Law* (Oxford University Press, 2016) 202, 204; Joseph Raz, *The Authority of Law: Essays on Law and Morality* (Oxford University Press, 2009) 221; also compare Åke Frändberg, *From Rechtsstaat to Universal Law-State: An Essay in Philosophical Jurisprudence* (Springer, 2014) 52-56 who sees them as connected to autonomy and humanism.

62 See David Dyzenhaus, 'Process and Substance as Aspects of the Public Law Form' (2015) 74 *The Cambridge Law Journal* 284, 305; Raz (n 61) 212–223; and in the context of criminal law Petter Asp, Magnus Ulväng and Nils Jareborg, *Kriminalrättens Grunder* (Iustus, 2013) 46.

These understandings of qualitative legality may consequently be described as largely legal-internal in the sense that they revolve around legal/logical arguments such as the ability of law to govern behaviour, internal coherence, or legal certainty. Even to the extent the qualitative requirements have been labelled as 'moral', the morality in question has been described as an inner morality *of law*.[63] Normative legal theories ascribe qualitative legality moral value as it provides law with intrinsic qualities that help explain its authority.[64] In a different vein, opining that the requirements say nothing of the moral character of the aim the law is trying to achieve, but rather how well the law manages to convey and achieve this goal, certain legal positivists have instead described them as functional requirements.[65] In any case, the internal perspective of these theories is to a large extent intentional,[66] and not without merit, as the legal system as such is the object of study and the idea of this system being understood best from the inside has proven capable of generating valuable insights regarding the authority of law.

3.2 Qualitative legality in the practical adjudication of technology: the case of the European Court of Human Rights

The previously mentioned challenges to foreseeability will of course carry with them implications for legality from this internal understanding of the concept. It is, for example, difficult for the individual to ascertain the criteria that will assign her a certain threat score in the algorithmic calculus of police software. Saying nothing about whether this is necessarily the right thing to do, an individual who would prefer to conform to whatever ideal law enforcement would prefer, rather than be deemed a threat, will find that it may be very difficult to do so.[67] This is particularly so if the characteristics adding to a certain score are innate or impossible to alter; such as ethnicity, gender or the socioeconomic status of your parents. It may also be difficult for individuals to assert their due process rights when the use of a surveillance technology is secret or shrouded behind confidentiality agreements.[68] Finally, it is difficult for individuals to challenge privacy violations in courts when the use of a certain technology is known only to the law enforcement agency employing the method and there are no external parties involved.

While these challenges have not always been addressed directly by courts in the context of emerging technologies, there are ways in which qualitative legality can mitigate some of these concerns. To illustrate this, I will use the case law of the ECtHR and its continuous endeavour to uphold the protection of fundamental rights in the face of technological development.

Initially, it is worth noting that the court has held that only publicly available norms can fulfil the requirement of legality (expressed by the court as 'in accordance with the law').[69] Furthermore these norms must reach compatibility with the rule of law – including a certain level of clarity and foreseeability.[70] The application of this foreseeability in a technological context is, however, rather unclear, but it is likely that the ECtHR would approach the issue as one where individual foreseeability of potential consequences is the primary concern – which could imply a requirement of access to internal non-legislative material in order to understand the application of the rules.[71] In contexts such as secret surveillance, where foreseeability cannot reasonably be construed as a possibility for an individual to foresee precisely when the authorities are likely to intercept his or her communication, the concern is instead one of limited discretion of government agencies.[72] In a technological context where there is a potential interference with a convention right such as the right to private life, this will necessitate that government agencies are not given a *carte blanche* to implement any technology they see fit, as doing so would increase discretion to the point of arbitrariness, potentially bypassing existing legal safeguards and failing to meet the standard of legality.[73]

The ECtHR has also been clear that any development in the interpretation of surveillance mandates because of technological development must be foreseeable to individuals through clear and accessible developments in case law. This maintains individual foreseeability when new technology, such as the Global Positioning System (GPS) trackers in the case of *Uzun v. Germany*, are applied, while avoiding legislation that is rigid and unable to handle technological developments that can be contained within a reasonable interpretation of the language of the law.[74] Furthermore, the ECtHR has quite consistently regarded new technologies in light of the safeguards around which their application is surrounded. In the case of GPS trackers, for instance, the court took note of the continuous review by German courts which had the power to disallow evidence.[75] In other cases where safeguards have been lacking, the court has been less inclined to accept surveillance measures.[76]

While cases relating to risk profiling have been rare in the ECtHR jurisprudence so far, the case of *Ivashchenko v. Russia*, regarding the copying of data from a laptop during border controls in Russia, gave the court an opportunity to begin approaching the issue. In this case, the court explicitly dismissed the notion that a risk-profiling approach applied by domestic authorities could be seen as a safeguard against arbitrary interference, when the application of this approach in regards to a specific individual would not be specified.[77] This case may indicate that a wide mandate in law cannot be cured by the application of narrower risk-assessment criteria set out in code. It also indicates that the use of risk-assessment profiles as support for coercive measures will need both a specific and foreseeable legislative basis and explainability in relation to the application of this profile to a certain individual.

63 See Fuller (n 56) 3–91. I will avoid the issue of morality here and use the term qualitative legality as it describes the function of the requirements without having to ascribe nor deny them such moral value.

64 David Dyzenhaus, 'Constitutionalism in an Old Key: Legality and Constituent Power' (2012) 1 *Global Constitutionalism* 229, 233.

65 HLA Hart, Essays in Jurisprudence and Philosophy (Clarendon Press, 1983); Raz (n 61) 226; compare also Marmor (n 56) 35–36.

66 See Dyzenhaus (n 64) 233.

67 Hildebrandt (n 1) 117.

68 See Joh (n 5) 39; Pell and Soghoian (n 24) 34–40.

69 *Leander v. Sweden* (1987) Series A No 116, § 54.

70 See *Huvig v. France* (1990) Series A No 176-B; *Kruslin v. France* (1990) Series A No 176-A.

71 See by analogy *Silver and Others v. the United Kingdom* (1983) Series A No 61, §§ 88-89, which concerned the screening of prisoners' letters, the detailed procedures of which was not set out in law but the prisoners concerned had been made 'sufficiently aware of their content', thereby surviving scrutiny under 'in accordance with the law'.

72 *Malone v. United Kingdom* (1984) Series A No 82. § 68.

73 See *Bykov v. Russia* App no 4378/02 (ECtHR, 10 March 2009) § 77–82, where the Russian legislation at the time allowed law enforcement authorities to conduct 'operative experiments' when investigating serious crime. This allowed unregulated surveillance technologies to be used, bypassing due process safeguards applicable to traditional communications surveillance.

74 See *Uzun v. Germany* ECHR 2010-VI, § 60–74.

75 *Uzun v. Germany* ECHR 2010-VI, § 69–74.

76 See *Ben Faiza v. France* App no 31446/12 (ECtHR, 8 February 2018); Liberty and Others v. the United Kingdom App no 58243/00 (ECtHR, 1 July 2008) § 62; Bykov v. Russia App no 4378/02 (ECtHR, 10 March 2009).

77 See *Ivashchenko v. Russia* App no 61064/10 (ECtHR, 13 February 2018) § 83.

Furthermore, states pioneering the implementation of emerging technologies will be subject to stricter scrutiny. As the court held in *S. and Marper v. the United Kingdom*, a case on the retention of DNA samples in the UK (as opposed to DNA profiles, which is common in other state parties to the convention) of persons no longer suspected or convicted of a crime:

> The Court observes that the protection afforded by Article 8 of the Convention would be unacceptably weakened if the use of modern scientific techniques in the criminal-justice system were allowed at any cost and without carefully balancing the potential benefits of the extensive use of such techniques against important private-life interests. In the Court's view, the strong consensus existing among the Contracting States in this respect is of considerable importance and narrows the margin of appreciation left to the respondent State in the assessment of the permissible limits of the interference with private life in this sphere. The Court considers that any State claiming a pioneer role in the development of new technologies bears special responsibility for striking the right balance in this regard.[78]

While this analysis by the ECtHR was made under the umbrella of proportionality, the court noted that the issue of legality in terms of legal safeguards is closely related to the analysis of proportionality.[79] The ruling of the court in the *Marper* case must be tempered by the sensitivity of the type of data involved. However, given the court's tendency to look at the consensus of signatory states to the ECHR when analysing an interference, the implication of claiming a pioneer role in terms of new technologies is likely to be applied in other cases.

This substantive approach to legality in the ECtHR case law has been combined with a dynamic approach to the possibility to lodge a complaint which is of relevance to opaque government measures. The regular approach of the court is to not review convention states' law and practice *in abstracto*, but instead to require individuals to show that they are directly affected by the measure at stake.[80] To allow a legal challenge against secret surveillance measures however, the ECtHR has adopted an increasingly generous approach to legal standing (victim status) under the convention. This was first established quite early on in the case of Klass and others v. Germany from 1978, where the court found that the mere existence of secret surveillance measures combined with the importance of ensuring effective control and supervision of them could warrant exceptions to the main rule.[81] The situations where such an approach could be warranted would, according to the ECtHR, have to be determined on a case-by-case basis.[82] As elaborated in the more recent case of *Kennedy v. United Kingdom*, the principle reason for this departure from its general approach 'was to ensure that the secrecy of such measures did not result in the measures being effectively unchallengeable and outside the supervision of the national judicial authorities and the Court'.[83] This line of reasoning has recently been extensively articulated in the case of *Roman Zakharov v. Russia*. Here, the ECtHR took account first of the scope of the legislation permitting secret surveillance measures and the potential of an applicant being affected by it, and secondly the available remedies on the national level and the effectiveness of those remedies. When there is suspicion and concern among the general public that secret surveillance powers are being abused, those concerns cannot be said to be unjustified in light of weak domestic remedies.[84]

The availability of legal safeguards overlaps with legality not only in the sense that lacking safeguards may result in arbitrary powers as in the cases mentioned above. Giving wide discretionary powers to authorities can result in a situation where individuals face great obstacles in trying to show before national courts that the actions of government authorities have been unlawful or unjustified. The resulting lack of meaningful court review in such cases may in itself create possibilities of abuse or arbitrariness which the court has found problematic.[85]

The approach by the ECtHR in surveillance cases has been interpreted as a sign of the court adopting a republican 'non-domination principle', where the effects of law on the power relationship between the state and citizen are taken into account when analysing the potential violation of a right under the convention.[86] Such an approach could potentially assist the ECtHR in navigating the more abstract and opaque interferences that new technologies such as big data and algorithmic decision-making might cause. A similar rise in non-domination conceptions of privacy and the impact of new technologies has been identified in the case law of the Court of Justice of the European Union (CJEU), where it has been linked to the need to restrict the accumulation of arbitrary powers.[87]

These developments in the case law of the ECtHR and the CJEU, while far from offering a comprehensive approach to new technologies, may help maintain legality in the sense of individual foreseeability. It provides a minimum level of transparency and foreseeability of government measures that may be applied to technologically mediated government and could help individuals challenge certain opaque measures. However, in approaching these issues, it is important not to lose track of the role that qualitative legality plays in a larger constitutional framework – extending beyond the individual to the democratic core of the state. This role will be further analysed in the following section and it will eventually give us a reason to return to, and elaborate on, the principles drawn up by the ECtHR.

4. Legality and democracy: dusting off Implicit interconnections

The theoretical outline I have previously presented of the concept of qualitative legality has largely been focused on legal certainty and foreseeability for individuals and the preservation of internal legal rationality. However, the maintenance of foreseeable legislation in the face of technologically mediated governing also carries with it important implications for democracy which will here be analysed further.

Assuming we base our understanding of legal legitimacy on the fulfilment of rule of law ideals, it follows from the implications to foreseeability that technologically mediated governing risks undermining the

78 *S. and Marper v. the United Kingdom* ECHR [GC], 2008-V, § 112. See also *Aycaguer v. France* App no 8806/12 (ECtHR 22 June 2017).
79 *S. and Marper v. the United Kingdom* ECHR [GC], 2008-V, § 98.
80 See *N.C. v. Italy* ECHR [GC] 2002X, § 56,; *Krone Verlag GmbH & Co. KG v. Austria* (no. 4) App no 72331/01 (ECHR, 9 November 2006) § 26; *Centre for Legal Resources on behalf of Valentin Câmpeanu v. Romania* ECHR [GC] 2014-V, § 101.
81 *Klass and others v. Germany* (1978) Series A No 28, § 34.
82 *Klass and others v. Germany* (1978) Series A No 28, § 34.
83 *Kennedy v. the United Kingdom* App no 26839/05) (ECtHR 18 May 2010) § 124.
84 *Roman Zakharov v. Russia* ECHR [GC], 2015-VIII, § 171.
85 *Ivashchenko v. Russia* App no 61064/10 (ECtHR, 13 February 2018) §§ 88-92.
86 Bart van der Sloot, 'A New Approach to the Right to Privacy, or How the European Court of Human Rights Embraced the Non-Domination Principle' (2018) 34 *Computer Law & Security Review* 539.
87 See Andrew Roberts, 'Privacy, Data Retention and Domination: Digital Rights Ireland Ltd v Minister for Communications: Privacy, Data Retention and Domination' (2015) 78 *The Modern Law Review* 535.

legitimacy of legal rules. This intra-legal legitimacy is however implicitly tied to broader issues of democratic legitimacy. The foundation of this democratic legitimacy can be sought in different sources. I will proceed with a conceptualisation of *democratic* legitimacy inspired by consent theories and the theory of deliberative democracy articulated by Jürgen Habermas. While I acknowledge that this is a concept that carries with it a somewhat thicker understanding of the 'oughts' of democratic processes, I believe it is one that resonates with most European democracies as a hybrid of liberal and republican values.[88] The implications I point to will in any case prove relevant in constitutional contexts where parliament carries the core of the democratic grounding of state power and where the separation of power is functionally important.

4.1 Qualitative legality as catalyst of deliberation

The idea of parliament as a democratic shorthand for 'the will of the people' is based on a presumption of the democratic nature of parliamentary law making. This democratic nature has its basis in both the direct nature of parliamentary elections, and in parliament as a place for democratic discourse and debate.[89] In its ideal form, the legislative process will subject bills to scrutiny and deliberation, and through this process parliament will both increase the quality of the bill through rational argumentation and ensure that their content can gain a majority support by the representatives of the public.[90] While doing so, the elected will be subject to pressure from the public and interest organisations, and to scrutiny by the media, ideally fulfilling the role of bringing issues from the periphery into the centre of public and political discourse. Meanwhile, on a political level, parliamentarians are subject to pressure from their party, the executive branch, and their primary constituents.[91] As emphasised by Habermas, the resulting discourse is the foundation of democratic legitimacy. It also implies something else. By ensuring that law is the result of a transparent democratic discourse, citizens – ideally – can see themselves as co-authors of the law they are subject to.[92]

Qualitative legality, as discussed above, can strengthen this democratic deliberation in several ways. By converting the political goals of the elected into legal norms, public policy is given a shape that allows legal-internal rationality and rule of law values to be upheld and makes politics legally enforceable.[93] As pointed out by Dyzenhaus, law's claim of authority must be understood as an implicit claim of legitimate authority, where legitimacy is dependent on legality as a rule of law value, creating the preconditions for a genuine social contract and consent, and where the subjects of the law are autonomous and partners in the rule of law project.[94] Qualitative legality thus ensures a connection between the law and the interests of the people as expressed through the elected legislative assembly's political deliberations and decisions. Indeed, the clearer the connection is between the actions of the state and the concrete legal form of the political decisions resulting from the deliberative and reflective process of representative democracy, the more democratic legitimacy. Maintaining qualitative requirements of legality will uphold a vital link between the language of the law (which holds democratic legitimacy through the deliberations and decisions that precede it) and the actions and decisions of the state.

The logic behind this argument becomes clearer if we think about foreseeability as something having effects in two directions. On the one hand, the individual is supposed to be able to foresee the effects of law on her actions, but this is practically impossible if legislators are not able to foresee the effects of law as mediated through technology. As these practices drift from what the legislator explicitly or implicitly could have foreseen, the legislation loses connection with the deliberative processes of democracy that underpin its legitimacy.[95]

The need to reach qualitative legality requirements further serves to create and increase transparency, enabling the democratic discourse surrounding current or proposed laws to be based on reasonable levels of foreseeability regarding the potential effects of those laws in relation to, for example, the impact on constitutional rights. Conversely, deficiencies in qualitative legality may result in a situation where neither citizens nor elected legislators really understand the implications of a proposed law, nor the power it confers to the executive. This is especially important when the legal practice is opaque or secret. One poignant example of this can be found in the United States where neither legislators nor citizens seemed able to foresee the vast surveillance system enabled by a vague section of the *USA Patriot Act* and the powers the executive government would eventually carve out of it.[96] While the (unintended) visibility of these surveillance practices through the disclosures of Edward Snowden did not lead to their discontinuation, it did contribute to the democratic debate on the security services' methods being based on a higher degree of foreseeability into the actual effects of the legislative framework and the actions of government agencies.[97] It has also allowed citizens to show standing to challenge the legality of the surveillance regime and the participation of their governments in it.[98]

In this context, qualitative legality serves an additional important function. It serves to uphold the separation of powers by limiting the discretionary power of the executive, while also upholding legal certainty by requiring that individuals can foresee what law requires and the authority given over them to executive agencies. This is mirrored in that clarity with regards to effects and powers conferred enables the elected representatives to foresee the scope of the power

88 For this interpretation of Habermas, see Lasse Thomassen, *Habermas: A Guide for the Perplexed* (Continuum, 2010) 121.

89 This idea recently received normative force in the German Federal Constitutional Court's decision on the European Financial Stability Facility and the ESM/Euro Plus Pact, where the court held that the Bundestag's right to decide the budget have to be exercised through deliberation and decision-making in the plenary setting rather than delegated to a committee or to the executive or a supranational mechanism, see Tony Prosser, 'Constitutions as Communication' (2017) 15 *International Journal of Constitutional Law* 1039, 1061.

90 See Jürgen Habermas, *Between Facts and Norms: Contributions to a Discourse Theory of Law and Democracy* (Polity, 1996) 304–306.

91 Antje von Ungern-Sternberg, 'German Federal Constitutional Court Parliaments — Fig Leaf or Heartbeat of Democracy? Judgment of 7 September 2011, Euro Rescue Package' (2012) 8 *European Constitutional Law Review* 304, 320-321; See also Prosser (n 89) 1059–1061.

92 Habermas (n 90) 449.

93 Dyzenhaus (n 62) 297.

94 Dyzenhaus (n 64) 259.

95 Cf. Habermas (n 90) 450; see also Tamanaha (n 55) 99-100. In the same vein, the connection to the proportionality assessments made by the legislator becomes less pronounced as well, indicating the need for a stricter review by courts. This is however a matter for a different discussion.

96 See Jim Sensenbrenner, 'This Abuse of the Patriot Act Must End | Jim Sensenbrenner' *The Guardian* (9 June 2013) https://www.theguardian.com/commentisfree/2013/jun/09/abuse-patriot-act-must-end accessed 14 August 2019.

97 Illustrative in this context are the investigations by the German Bundestag – the 'Untersuchungsausschuss "NSA"' – and the EU parliament investigations 'The US surveillance programmes and their impact on EU citizens' fundamental rights' (PE 474.405) and 'Human rights and technology: the impact of intrusion and surveillance systems on human rights in third countries' (2014/2232(INI)).

98 See *Big Brother Watch and others v. the United Kingdom* App no 58170/13 (ECtHR, 13 September 2018).

handed to the executive and enables an informed democratic debate and discussion on such transferals of power based on a reasonably foreseeable practice.

While qualitative legality can fulfil these democratic functions, it is challenged when the actual effect of law is mediated through technology in ways that impacts foreseeability. In such cases, democratic discourse might be based on limited information and with a diffuse conception of the actual implications of laws under deliberation. It may result in a situation where legislators cannot reasonably foresee the implications of a law or the powers it confers to the executive government, either through a wide interpretive space, or through technological developments that carve out further power from discretionary spaces over time. It may also allow governments to hide or disguise the exercise of power, by clouding them in code. As Lessig puts it, '[c]ode-based regulation – especially of people who are not themselves technically expert – risks making regulation invisible.'[99] He argues that transparency serves as an important check on government power and the only rules government power should impose are those that would be obeyed if imposed transparently.[100]

In the context of technology, these transparent deliberative practices are sometimes described as difficult to achieve due to a perceived inability of the public to navigate complex technological issues that arise in many policy areas. This has led to a questioning of the ideal of deliberative democracy in this context.[101] More recent research in science and technology studies (STS) has however challenged this assumed ignorance and explored instead the differing points of departure from which people navigate and question technological choices and social dilemmas. These viewpoints and experiences may be different from those of experts and politicians, but equally valid and complementary, highlighting the need to maintain public deliberation of emerging technologies and their implications.[102] In any case, it is fairly obvious that navigating complex social and technological issues is not made easier by keeping them opaque and vague. In fact, a more transparent democratic deliberation can assist governments in achieving compliance with human rights norms, avoiding issues with both legality and proportionality.[103]

An example from Sweden illustrates this. In 2007 a government bill intended to give the Swedish signals intelligence agency (FRA) access to all wired network traffic crossing Swedish borders to allow for automated searches for combinations of keywords and characteristics deemed relevant for national security, so called 'massive interception' or 'bulk collection'.[104] This led to significant public debate and parliamentary infighting, even within the ruling coalition government, over fears of mass surveillance. To pass the bill, the government announced proposals to strengthen the oversight mechanisms, adding – among other things – a court review of search terms and limiting access to only those fibre optic information carriers which are likely to be relevant for the particular intelligence target.[105] These changes were recently highlighted by a chamber judgment of the ECtHR as key aspects making the law acceptable under the ECHR.[106]

Consequently, the public and parliamentary debate did not only force the government to increase oversight, it also forced it to articulate rather vague and potentially wide legislative language into something more specific and transparent that ultimately managed to gain support in parliament and which may yet survive scrutiny by the ECtHR.[107] This, in turn, served to limit the discretionary space of the signals intelligence agency, maintaining a minimum level of legality, while simultaneously calming the concerns from parliamentarians and certain sections of the public.

This ability of deliberative practices to 'act as a prophylactic against later costly lawsuits'[108] is often forgotten. In a constitutional context it can also reduce the risk of legal uncertainties because of legislation that is expensive to implement, yet cannot for long be applied or is simply declared invalid following a decision by a court.[109]

4.2 Allowing autonomy and fostering cross institutional discourse

So far, I have touched upon the role qualitative legality plays for the legislative process. There are, however, further functions that qualitative legality can fulfil to allow for a broader deliberative discourse in a democratic state.

As mentioned above, democracy is more than a simple expression of popular will, it is grounded in a *process*. As conceptualised by Habermas through his co-originality thesis, any democratic system must capture both public and private autonomy, ensuring that citizens have a standing to both express a political will and assert their constitutional rights.[110] In this process, courts are tasked with the important role of interpreting and applying law as well as acting as guardians of individual rights. As the ECtHR has concluded, a gradual and foreseeable development of law through legal precedent is not incompatible with qualitative legality.[111] As far as interpretation of legislative acts goes, there is a point however, where the connecting strands between a legally authoritative interpretation which is foreseeable due to gradual developments in case law, on the one hand, and the democratic legitimacy of parliament on the other, is severed. The question then becomes if the legal system can cure a lacking *ex ante* democratic deliberation regarding a specific technological reality with *ex post* judicial means of maintaining individual autonomy? If we adopt a wider understanding of how and when the deliberative practices can be realised, we can reasonably include not only the ability of citizens to engage in public discourse in advance of legislative measures being put in place, but also the way citizens may challenge the constitutionality of law and government measures in courts, asserting their autonomy as legal subjects and actors within a constitutional framework. In doing so, they can bring constitutional issues under the purview

99 Lessig (n 14) 138.
100 Lessig (n 14) 328.
101 See Patrick W Hamlett, 'Technology Theory and Deliberative Democracy' (2003) 28 *Science, Technology, & Human Values* 112, p. 125.
102 See Peter Newell, 'Democratising Biotechnology? Deliberation, Participation and Social Regulation in a Neo-Liberal World' (2010) 36 *Review of International Studies* 471, 477–478, discussing the context of environmental risk and GMO.
103 Hamlett (n 101) 130.
104 Swedish Government Bill [2006/07:63].
105 Swedish Government Bill [2008/09:201].
106 See *Centrum för Rättvisa v. Sweden* App no 35252/08 (ECtHR, 19 June 2018) § 180.
107 The case has recently been referred to the Grand Chamber of the ECtHR.
108 Hamlett (n 101) 130.
109 The invalidation of the EU data retention directive and the subsequent rejection of its national implementation law in Sweden is a poignant reminder of this, see Joint cases C-293/12 & C-594/12, *Digital Rights Ireland Ltd v. Minister for Communications, Marine and Natural Resources and others, and Kärntner Landesregierung and others*, EU:C:2014:238; Joined Cases C-203/15 and C-698/15, *Tele2 Sverige AB v Post- och telestyrelsen and Secretary of State for the Home Department v Tom Watson and Others* (2016), Judgment of the Court (Grand Chamber) of 21 December 2016, ECLI:EU:C:2016:970.
110 Habermas (n 90) 121–123.
111 See section 3.2 above.

of courts – essentially activating a constitutional discourse between courts, government, and parliament.[112]

This control will in turn enable the autonomy and dignity of the individual to be safeguarded, and as such the preconditions for both the formation of public opinion, the expression of this opinion, and the retention of a democratic system that allows individuals to authorise future legislative assemblies to act on their behalf.[113] To enable this, courts must however be open to a more generous approach to standing, as the sometimes subtle *individual* effects can mask more overarching *systemic* issues. In this sense, the ECtHR with its dynamic approach to victim status in surveillance cases can be one example of how to balance the interests involved.[114]

I believe this perspective is a fruitful addition to the concept of republican non-domination as it is connected to similar ideas – distribution of power, our relationship as citizens with government bureaucracies, and the avoidance of discretionary power.[115] It also engages similar issues of democratic inclusion as a counteraction to domination.[116] But it also engages with further questions of power transferrals between government branches, the existence of deliberation regarding the application of a specific technology, as well as the possibility for individuals to assert themselves as autonomous legal actors through the courts.

This brings us to the question of how legality may be understood to safeguard both individual and democratic functions in the light of technologically mediated governing and black box policing. Or, in other words, how should legality be recoded to fit within a technological legal framework?

5. Qualitative legality recoded

The central issue that this contribution has so far orbited (albeit in a rather twisted trajectory) is how technologically mediated governing – particularly in the policing context – can be legally contained and regulated, and how legality in the context of such governing can be upheld in adjudication. So far, I have primarily focused on certain challenges relating to technologically mediated governing and pointed to some tentative responses to those challenges from the ECtHR. I have also outlined the legal and democratic functions that legality fulfils and in doing so attempted to highlight the values that regulation and adjudication in this context should try to uphold. In the following, a more constructive approach, with every intellectual peril that entails, will be attempted.

5.1 Technology neutrality

While academics have, as is evident above, pointed to the risks involved in allowing new technologies to run rampant through the regulatory environment of policing, the response from legislators has often been considerably more innovation-friendly. This is especially evident through the concept of technology neutrality that has been a staple of technological regulation in both the EU and the US since the 1990s.[117] As put by Reed, the idea that law should not pinpoint a certain technology, but rather keep itself open to technological development by remaining technology neutral, has been regarded as naturally good, 'like motherhood and apple pie'.[118] This ideal, however, is likely to exacerbate the very issues highlighted here. The point of technology neutral law is often to allow authorities to choose suitable technologies to achieve a government policy, thereby avoiding rigid or outdated legislation. To achieve this, purposes or generalised technological concepts are described to avoid specific references to technology which may become outdated. However, qualitative legality as a concept would (anthropomorphically) frown upon precisely this form of discretion. Not only does it create uncertainty as to how the law is to be interpreted in relation to emerging technology, but the technological affordances that were the point of departure for the deliberations in the legislature may fundamentally shift. The gradual adaptation to new technology that technology neutrality was supposed to ensure, may instead create wide discretionary areas of technologically mediated governing. The risk, essentially, is a transferral of power from parliament (choosing to open up the discretionary technological space) to the executive agencies implementing a certain technology which, depending on the context, may never be subject to review by a court. The black box of policing discussed above is, in other words, nourished by the apple pie of technological neutrality.

In this context, it is worth keeping in mind that there are two distinct types of neutrality. First, there is a very reasonable ideal that constitutional rules and principles should be insusceptible to technological change. As Lessig puts it, judges are translators:

> We must always adopt readings of the Constitution that preserve its original values. When dealing with cyberspace, judges are to be translators: Different technologies are the different languages, and the aim is to find a reading of the Constitution that preserves its meaning from one world's technology to another.[119]

In terms of the second type of neutrality, which provides government agencies with mandates to exercise power through legislation that does not specify technological means, there is however a risk that technology neutral legislation instead codifies a form of indifference to the importance of code and architecture. It becomes in effect a transferal of power from the democratic arena to the architects of the digital arena; in some cases, this shifts power from the state to markets, in others from parliament to government agencies. In many cases it is both.

It is worth considering that the requirements of qualitative legality, including the deliberative aspects I have argued for above, may demand a more specific legislation – at least in such legislative contexts that may affect individual rights or the power relationship between citizen and state. The need to revisit legislation more frequently in view of new technological developments, while understandably a complicated and time-consuming process, may be a worthwhile price to pay to foster both legality and democratic legitimacy in the technological context.

5.2 A more extensive interpretation of legality

To counter unconstrained transferrals of power, we need to understand the implications of technology, not just in terms of certain individuals or groups at risk of suffering adverse effects, but also the shifts in the power relationship between individuals and the govern-

112 Prosser (n 89) 1059–1061.

113 Cf. Mattias Kumm, 'Democracy Is Not Enough: Rights, Proportionality and the Point of Judicial Review' (Social Science Research Network 2009) SSRN Scholarly Paper ID 1356793 https://papers.ssrn.com/abstract=1356793 accessed 14 August 2019.

114 See section 3.2 above.

115 See Andrew Roberts, 'Forewords · Why Privacy and Domination?' (2018) 4 *European Data Protection Law Review* 5.

116 Ludvig Beckman and Jonas Hultin Rosenberg, 'Freedom as Non-Domination and Democratic Inclusion' (2018) 24 *Res Publica* 181.

117 See generally Chris Reed, 'Taking Sides on Technology Neutrality' (2007) 4 *Script-ed* 263; Paul Ohm, 'The Argument against Technology-Neutral Surveillance Laws' (2010) 88 *Texas Law Review* 1685.

118 Reed (n 117) 264-265.

119 Lessig (n 14) 165-166.

ment. Conceptualising the legality of new technologies must therefore go beyond 'due process legality', focusing on the particular effects of an individual, and *also* ask wider questions regarding the transferral of power from law – the purvey of parliament – to the technologically mediated bureaucracies of executive agencies and the private technology companies they rely on.

As courts analyse the legality of a certain measure, they should consider the potentialities of technology to shift power relationships within the branches of government and between state and private actors. In doing so, even within the limits of a single case, courts may need to consider the wider implications of a certain technology and whether they are transparent and foreseeable not only for the individual, but also whether they were ever the subject of democratic deliberation at all.

Admittedly, extending the analysis of legality beyond the case at hand might extend the purview of the court into what some may believe would amount to judicial activism, and the counterargument may be that courts should instead defer to the government if in doubt. I would however argue that when the legislator has not even considered the use of a certain technology, there is no legislative will of parliament to defer to.[120] Deferring to the government in such cases would instead cause an implicit transferral of power from parliament to the executive that was never intended. In contrast, by keeping in mind the democratic functions of qualitative legality and strictly analysing the legality of a technological measure in that light, courts instead serve parliamentary supremacy by essentially turning the question back to the proper place for democratic deliberation. In doing so, courts will essentially say; 'if this is what parliament desires, it will at least have say so explicitly, transparently, and after deliberating on the issue'.[121]

5.3 Judicial pre-review and extensive *ex post* review

One way in which the application of new technologies could be better insured against a departure from the requirements of legality, while maintaining some flexibility, is through preliminary court reviews of new technologies being implemented within public agencies that may affect individual rights or due process.

While such reviews of legality are often conducted within executive agencies prior to the application of certain methods or technologies, the addition of a court review could fulfil functions that improve qualitative legality in several ways. Following an internal review of the legislative framework surrounding a new or previously untested method, a law enforcement agency could apply to a court to get a preliminary approval of its use, making their best arguments for why it may be legal. This hypothetical court review could then consider both how well the new technology fulfils existing requirements of legality and proportionality, as well as its fit within legal mandates and due process requirements. Simultaneously, civil society organisations, bar associations, and other stakeholders could file their own briefs to inform the court. Should the method not fit within the existing legal and deliberative framework (i.e. considering the degree to which the method could have been foreseen and deliberated within the democratic process), the court could refer the issue to the legislature. Should it fit, but with certain caveats, the court could put in place such terms and conditions that are required to limit the use of the technology to what is allowed within the framework of constitutional or human rights rules. Such a preliminary review could also make relevant legal aspects of the application of the method public, reaching the transparency and foreseeability requirements similar to the case of *Uzun v. Germany* mentioned in section 3.2 above.

The use of preliminary review of specific technologies is different from other measures such as judicial review *in abstracto*, as it focuses not on the legal rules themselves, but instead on the technologies used and how they fit within a legal framework. Comparable solutions implemented within a political framework exist in certain US cities, most famously in Seattle, where a city surveillance ordinance requires the police to report the use of surveillance technologies onto a 'master list' which is then subject to public deliberation and city council review. It is intended to increase political control of surveillance technologies and to increase civil society involvement while increasing public trust in the police.[122] While the Seattle ordinance has been seen to not properly address the use of algorithmic surveillance,[123] it is still a noteworthy example of how technologies can be subjected to increased scrutiny.

While the publication of details of surveillance methods is – to put it mildly – frowned upon by intelligence and law enforcement agencies, the clarification of more general attributes of surveillance mandates (such as the general scope of its intrusion into a right and the safeguards surrounding it) and the relevant legal aspects of how a technology can be reconciled under a legal mandate, are in any case of the type that needs to be publicly available to reach legality requirements (as they are construed by the ECtHR).

To avoid the negative effects of technology neutrality discussed above, a pre-review should strive to delineate the salient features and underlying presumptions that distinguish the legal analysis of the method in terms of impact on individual rights, principles, or rules. This will ensure that shifts in technologies impacting those underlying features and assumptions will necessitate a new review. In relation to surveillance technologies, this could imply a description of the limits in terms of the degree to which the method allows for the mapping of individuals or groups. In relation to the implementation of machine-learning algorithms, this could imply a description of the necessary level of human involvement in decision-making, restrictions on allowed applications, attributes or inferences, restrictions in the further measures taken based on automated profiles, necessary measures to quality assure underlying data sets, or safeguards in terms of *ex post* auditing.

It is important to note that a review, such as the one outlined above, can only ever be preliminary and must not be allowed to prevent a later *ex post* judicial review of the application of the technology used. As discussed in previous sections, the actual effects of a certain technology are in many ways dependent on its application and its interface with citizens. The preliminary review can, however, ensure a legal check on otherwise discrete and direct technological measures. It would also serve as a continuously updated inventory of techno-

120 This conclusion is inspired by that of the ECtHR judge Robert Spano, opining that deference to national parliaments in questions of proportionality is not a valid argument in the cases where the national parliament has never considered the proportionality in the first place. Robert Spano, 'The European Court of Human Rights and National Courts: A Constructive Conversation or a Dialogue of Disrespect?' (2015) 33 *Nordic Journal of Human Rights* 1, 7.

121 See further Markus Naarttijärvi, 'Kvalitativ Legalitet: Ett Demokratiskt Perspektiv' (2018) 131 *Tidsskrift for Rettsvitenskap* 206, 206-234.

122 See Meg Young, Michael Katell and PM Krafft, 'Municipal Surveillance Regulation and Algorithmic Accountability' (2019) 6 *Big Data & Society* 205395171986849. Similar ordinances exist in Berkeley, Cambridge, Davis, Nashville, and Oakland.

123 Young et al (n 122) 12.

logical methods and measures developed or applied within government agencies, increasing transparency. Even if certain aspects or methods would need to be kept under a shroud of secrecy, access to these decisions by oversight organs, researchers, and parliamentary committees would inform the legislative process in the technological context.

5.4 *Ex post* auditing

The importance of algorithms and the data that fuel them is becoming increasingly clear, and there have been increased efforts to ensure some insight into algorithms. In Europe, this push has not been fuelled by concerns of legality, but rather from the viewpoint of data protection, privacy, and informational self-determination. Within the European Union, steps have been taken to try to achieve transparency and limit the impact of profiling and algorithmic decision-making through legislation such as the new EU General Data Protection Regulation (GDPR).[124] Article 13.2(f) GDPR specifically requires the provision of meaningful information about the logic involved in automated decision-making and profiling, as well as the significance and the envisaged consequences of such processing of personal data for the data subject. There is a further rule in article 22, giving data subjects a right not to be subject to decisions made *solely* on automated processing, including profiling, which produces legal effects for him or her or significantly affects him or her. However, the impact of this rule is limited in two primary ways. First, the rule only applies to *fully* automated decision-making – including a human in some part of the decision-making process will circumvent the rule as long as the human has meaningful impact on the outcome.[125] Second, the GDPR does not apply to processing of personal data within law enforcement and while there is a similar rule in the directive harmonising data protection in that context, it is possible for member states to allow such automated decision-making through national law though not based on certain sensitive categories of data.[126]

Still, the impulse to ensure access to and information about algorithmic decisions based on citizen data is reasonable. Even when the applications of technology are in accordance with the law, transparency can create awareness of how data are used and how government agencies (and, in the case of the GDPR – even private actors) reach their conclusions based on these data. It is however difficult to achieve full transparency, both on account of the technologies involved such as neural networks where the logic may make review fruitless, and because true understanding of the outcomes will require access to the underlying data, which may end up conflicting with privacy and data protection of others whose data are being processed. There may therefore be an increased need for expert auditing of big data governing from oversight bodies where access to information and technological experts can be achieved more effectively.[127] Importantly, the organs auditing such technology should be given mandates which are not tied to express technologies or policing powers as this runs the risk of new technologies being implemented in the gaps between these mandates. Instead, their auditing mandates should be wide and overarching to allow their audit to adapt to changing circumstances.

5.5 Avoiding determinism

The advent of technologically mediated governing does not entail a necessary surrender of legal values to the unrelenting march of technological development. Technology challenges the existing framework of legal governance and involves inevitable difficulties in regulating technology. However, as noted in STS literature, the surrender to technological determinism through the idea that technological change causes or determines social change 'leaves no space for human choice or intervention and, moreover, absolves us from responsibility for the technologies we make and use'.[128] In fact, there is nothing forcing government agencies to employ technological measures or make governing dependent on their application. Indeed, while technological determinism is often visible in the debates on regulating social media, drones, or AI for private entities, the normative influence of law within government entities is, or at least should be, higher. As such, even while we may accept the difficulty of effectively preventing a certain technology from affecting the everyday life-world of private citizens or private entities, this does not answer the question of whether we should allow or pursue the use of the same technology within our government agencies. Instead, these are choices governments can make and abandoning these choices to the whims of technological trends will fundamentally weaken the sphere of democratic deliberation. As Lessig puts it: 'Code codifies values, and yet, oddly, most people speak as if code were just a question of engineering. Or as if code is best left to the market. Or best left unaddressed by government.'[129]

Avoiding this determinism requires us to 'recode' legality to fit a technological context. Doing so will essentially require three main considerations to be actively acknowledged in both the legislative process and the adjudication of technologically mediated governing.

First, as I have pointed out above, the legislative process must be based on a reasonable level of foreseeability regarding the interaction between law and technology. This may require the abandonment of technology neutrality as a legislative ideal in contexts where technology will interfere with rights, alters the power relationship between citizen and state, or when it significantly affects the balance of power within a constitutional system. If government power should be bound by law, technology cannot be exempt from this.

Second, the review of legality of technological measures by courts should consider the existence of deliberative practices underpinning

124 Regulation (EU) 2016/679 of the European Parliament and of the Council of 27 April 2016 on the protection of natural persons with regard to the processing of personal data and on the free movement of such data, and repealing Directive 95/46/EC (General Data Protection Regulation).

125 As put by the Article 29 Working Party: 'The controller cannot avoid the Article 22 provisions by fabricating human involvement. For example, if someone routinely applies automatically generated profiles to individuals without any actual influence on the result, this would still be a decision based solely on automated processing. To qualify as human involvement, the controller must ensure that any oversight of the decision is meaningful, rather than just a token gesture. It should be carried out by someone who has the authority and competence to change the decision. As part of the analysis, they should consider all the relevant data.' See Article 29 Working Party, '*Guidelines on Automated Individual Decision-Making and Profiling for the Purposes of Regulation 2016/679*' (Article 29 Working Party 2018) wp251rev.01, 21.

126 Article 11, Directive (EU) 2016/680 of the European Parliament and of the Council of 27 April 2016 on the protection of natural persons with regard to the processing of personal data by competent authorities for the purposes of the prevention, investigation, detection or prosecution of criminal offences or the execution of criminal penalties, and on the free movement of such data, and repealing Council Framework Decision 2008/977/JHA

127 See Paul B de Laat, 'Algorithmic Decision-Making Based on Machine Learning from Big Data: Can Transparency Restore Accountability?' (2018) 31 *Philosophy & Technology* 525.

128 Sally Wyatt, 'Technological Determinism Is Dead; Long Live Technological Determinism' in Edward J Hackett and others (eds), *The Handbook of Science and Technology Studies* (3rd edn, MIT Press, 2008) 169.

129 Lessig (n 14) 78.

the measure under review. While a certain technology may fit into the semantic meaning of a legal provision, the effects produced may never have been possible for legislators to envision. While the point here is not that every consequence of technology must have been foreseen – which would make law unbearably complex and rigid – measures that will substantially impact rights or the power relationship between citizen and state, or parliament and the executive, should be subject to a stricter review.

Third and finally, the many subtle ways in which technologically mediated governing can influence individuals will require courts to have a dynamic and generous approach to standing. Here, the approach taken by the ECtHR can serve as inspiration. I have also suggested the implementation of a form of preliminary judicial review of new technologies that could assist in the fulfilment of qualitative legality in the application of emerging technologies in governing. In combination with a strict ex post court review and auditing by expert oversight bodies with access to both code and data, this could aid in the mitigation of the concerns raised here.

5.6 The choices we make

As we have seen, there are several important implications of technologically mediated governing for both legality as a rule of law value, and the implicit democratic values legality serves. This is true both in the context of policing and in other fields of governing. The pertinent question raised is whether automation of government decision-making will itself shape the rule of law.[130] If the development of the rule of law has made the exercise of government power subject to the law, increased foreseeability, and limited arbitrariness, we may indeed reasonably ask whether technologically mediated governing will move important aspects of this governing into a black box. In this box, the norms that govern are statistical rather than legal. The goal of foreseeability is replaced by ambitions of accuracy, and if human discretion is replaced, there is an inherent risk that it is replaced by an automated naivety regarding the systematic inequality which is represented in the data that surround us. Avoiding this will require us to interpret legality in a way that maintains both the explicit and implicit values it protects even in the face of technological change.

Acknowledgements

The author would like to thank the arrangers and participants of the *TILTing* 2019 conference, as well as the editors and anonymous reviewers of Technology and Regulation for valuable comments that helped inform and improve this paper. This contribution is a result of the project 'Policing in Sweden – Efficiency and Rule of Law in Police Work', the funding for which has generously been provided by Riksbankens Jubileumsfond (The Swedish Foundation for Humanities and Social Sciences), grant no. SGO14-1173:1. Parts of chapter 3 and 4 are built upon ideas the author has previously discussed in Swedish in Tidsskrift for Rettsvitenskap, vol. 131, 2–3/2018 p. 206–234, available at: https://www.idunn.no/tfr/2018/02-03/kvalitativ_legalitet

130 Monika Zalnieriute, Lyria Bennett Moses and George Williams, 'The Rule of Law and Automation of Government Decision-Making' (2019) 82 *The Modern Law Review* 425. See also Emre Bayamlıoglu and Ronald Leenes, 'The "Rule of Law" Implications of Data-Driven Decision-Making: A Techno-Regulatory Perspective' (2018) 10 *Law, Innovation and Technology* 295, 311.